BUILDING MAINTENANCE MANAGEMENT

Barrie Chanter
Principal Lecturer
Department of Building Surveying
De Montfort University, Leicester

Peter Swallow
Professor of Building Surveying
Department of Building Surveying
De Montfort University, Leicester

Blackwell
Science

Blackwell Science Ltd, a Blackwell Publishing
Company
Editorial Offices:
Osney Mead, Oxford OX2 0EL, UK
  Tel: +44 (0)1865 206206
Blackwell Science, Inc., 350 Main Street,
Malden, MA 02148-5018, USA
  Tel: +1 781 388 8250
Iowa State Press, a Blackwell Publishing Company,
2121 State Avenue, Ames, Iowa
50014-8300, USA
  Tel: +1 515 292 0140
Blackwell Publishing Asia Pty Ltd, 550 Swanston
Street, Carlton South, Melbourne, Victoria 3053,
Australia
  Tel: +61 (0)3 9347 0300
Blackwell Wissenschafts Verlag,
Kurfürstendamm 57, 10707 Berlin, Germany
  Tel: +49 (0)30 32 79 060

First published 1996
Reissued in paperback 2000
Reprinted 2001, 2003

Library of Congress
Cataloging-in-Publication Data
is available

ISBN 0-632-05766-1

A catalogue record for this title is available from
the British Library

Set in 10pt Times
by DP Photosetting, Aylesbury, Bucks
Printed and bound in India by Thomson
Press (I) Ltd.

For further information on
Blackwell Publishing, visit our website:
www.blackwellpublishing.com

Contents

Preface

The growth in the significance of building maintenance as a proportion of the output of the construction industry has taken place against a backdrop of mounting pressure on new-build activity, and a growing awareness of the need to manage the condition of the nation's building stock more effectively. Paralleling these developments has been the increased application of new technology, permitting the more efficient use of data. Notwithstanding this, it is still the case that much maintenance activity takes place in a context that does not create a fully integrated approach to managing building performance and, thus, the full potential of many buildings is never wholly realised.

There is a wealth of information and informed comment available concerning the poor condition of much of the UK's building stock and this formed a natural starting point for Chapter 1. This reviews some of the currently available information, supplemented by an overview of the nature and significance of the repair and maintenance sector, within the context of both the construction economy and the national economy as a whole.

The position of the maintenance department within the structure of the parent organisation is considered in Chapter 2 and confirms the existence of a major weakness concerning clients' perceptions of the importance of maintaining the building fabric. We are indebted here to several individuals and organisations for providing valuable insights, from the perspective of the professional maintenance practitioner, which enabled us to analyse a range of typical organisational structures, both as exemplars and as an indication of the diversity of thinking that exists. Our grateful thanks go to Stafford Taylor and his colleagues at Building Surveying Associates, together will all those others, who for reasons of confidentiality, it would not be appropriate to name here.

That poor detailed design affects building performance, and hence maintenance, is well known. Chapter 3 explores the design/maintenance relationship against the broader backcloth of the building procurement process. Many of the problems encountered in buildings stem from the brief development phase, where a failure to establish user requirements in sufficient detail results in the poor performance of the completed building. At the hand-over stage also there may be serious shortcomings, and more careful consideration must be given to providing the client with a proper building model to facilitate the effective management of the property. None of these crucial developments can take place without a major shift in client attitudes, and professions working within the built environment must shoulder some responsibility for this by a failure to educate clients sufficiently. For too long they have evaded a responsibility for adequately preparing uninformed clients, and the increasingly competitive pressures of competitive fee tendering are a serious obstacle in this respect.

The nature of maintenance work is examined in some detail in Chapter 4 because a full understanding of the types of maintenance work is necessary in order to appreciate the manner in which the maintenance workload is generated. However, this has to be tempered with a degree of realism and a recognition of the context within which it occurs.

A large quantity of data is generated by maintenance operations, and its management is a complex matter. Chapter 5, therefore, focuses specifically on information management and develops the notion of a building condition model. The technology to produce and maintain an electronic performance model now exists at a variety of levels of sophistication. A number of software packages are described to provide a flavour of current practice. We would like to thank many people for their help in this area: Steve Wilson of Minster General Housing Association; the staff of Building Surveying Associates; colleagues at De Montfort University notably, Donna Wilson, Tracey Burt, and Tony Gibbs in Property Services, Stuart Planner in Landstaff, Chris Watts of the Building Surveying Department, who provided invaluable guidance throughout the preparation of this chapter, and Rob Ashton of the CAD Centre who acted as an excellent specialist IT advisor. In developing the chapter on information management we were aware of two major factors. Firstly, the pace of software development is now so rapid that it is almost impossible to keep abreast of it, and thus it is only possible to provide a limited overview. Secondly, it was also apparent that in separating out maintenance planning into Chapter 6, there was some risk of duplication. However, it was felt that maintenance planning warranted separate consideration in order to clearly define good practice and principles. Chapters 5 and 6 therefore should be read together.

Chapter 7, which deals with maintenance contracts, logically precedes the consideration of maintenance execution, the subject of Chapter 8.

Inevitably there is some overlap from our consideration of detailed execution into a number of other areas which we recognise, but justify on the grounds that it is the context that is important.

Overlying the whole subject of maintenance management is the issue of facilities management. Throughout the text we have made reference to this rapidly developing discipline, particularly with respect to information management and the construction of a building performance model. Ultimately our real concern is with the performance of buildings, whether they are viewed as an investment, asset or facility in the widest sense. Perhaps the biggest service provided by the growth of facilities management has been to focus attention more sharply on, and to promote the profile and image of the people who manage buildings.

It is hoped that this book will act both as an update for practitioners and provide an introduction for final year students, graduates and others encountering building maintenance work for the first time.

In presenting this introduction we have thanked a number of people who have provided material help or acted as sounding boards for some of our ideas, but the biggest thank you of all must go to our respective wives, Lesley and Lynne, to whom we dedicate this book.

Barrie Chanter
Peter Swallow

List of Abbreviations

ACE	Association of Consulting Engineers
ADC	Association of District Councils
BEC	Building Employers Federation
BMCIS	Building Maintenance Cost Information Service
BMI	Building Maintenance Information Ltd
CA	contract administrator
CAD	computer aided draughting
CCPI	Co-ordinating Committee for Project Information
CDM	Construction (Design and Management) Regulations
CIOB	Chartered Institute of Building
CLASP	Consortium of Local Authorities Special Programme
DES	Department for Education and Science
DLO	direct labour organisation
DoE	Department of the Environment
DIY	do-it-yourself
EHCS	English House Condition Survey
FM	facilities management
GDP	gross domestic product
GNP	gross national product
HEC	Higher Education Corporation
IT	information technology
JCT	Joint Contracts Tribunal
LCC	life cycle cost
LEA	local education authority
NFHA	National Federation of Housing Associations
NEDO	National Economic Development Office
NHS	National Health Service
NPV	net present value
OMR	optical mark reader
PSA	Property Services Agency
RIBA	Royal Institute of British Architects

RICS	Royal Institution of Chartered Surveyors
RPI	retail price index
SO	supervising officer
SSHA	Scottish Special Housing Association

Chapter 1

The Maintenance Dimension

Building maintenance has consistently been treated as the 'poor relation' of the construction industry, attracting only a tacit recognition of its importance, both within the industry and amongst building owners. This manifests itself in a general lack of understanding of both its scope and significance by all parties to the building procurement, construction, and management processes. In consequence, the backlog of repair and maintenance work required to bring the country's building stock to a minimum acceptable level continues to grow at an unacceptable rate.

Latterly, for a number of reasons, the dimension of the problem has forced itself higher onto the agenda and promoted what appears to be greater professional interest. This increasing level of concern over the condition of the nation's building stock has served to expose more clearly the extent of the problem. Whilst effective maintenance policies are not by any means the norm, the efficient utilisation of scarce resources is beginning to be approached in a more informed way, and the fundamental relationship of the condition of a building's fabric to its total performance examined more critically.

Maintenance defined

BS 3811: 1984[1] defines maintenance as:

'A combination of any actions carried out to retain an item in, or restore it to an acceptable condition.'

From this definition two key components can be identified:

❏ actions that relate not only to the physical execution of maintenance work, but also those concerned with its initiation, financing and organisation
❏ the notion of an acceptable condition, which implies an under-

standing of the requirements for the effective usage of the building and its parts, which in turn compels broader consideration of building performance

The latter presents some problems when attempting to determine the standard which represents an acceptable condition, as opinions will vary from person to person and over time according to the type of building under consideration, its usage, and changing circumstances. Within the private sector, for example, there are likely to be quite different opinions as to what an acceptable condition is between parties on opposing sides of a tenancy agreement. These perceptions may range from a 'cosmetic' view to an in-depth evaluation of building performance needs, and will certainly be fluid enough to drift considerably with changing market conditions.

Within the public sector, it is apparent that social and political forces will have some bearing, and it may be extremely difficult to justify a given stance in the same terms that might be exercised in the private sector. The nature of the responsibilities of the public sector highlights the need to deal not only with the satisfaction of individuals, but groups of people and society in general.

What is an acceptable condition can also be looked at in quite another way by trying to view the physical needs of the building in an objective manner, although even then problems of perception will be just as prevalent. It is only necessary, for example, to contrast the different attitudes taken to fabric maintenance and the maintenance of engineering services to see this demonstrated. Acceptable standards will also change with time, and given the long life of buildings this is of major importance.

The Committee on Building Maintenance[2] recommended the adoption of the following definition of maintenance:

'... work undertaken in order to keep, restore or improve every facility, i.e. every part of the building, its services and surrounds to a currently acceptable standard and to sustain the utility and value of the facility.'

This definition is, if anything, rather broader than that in BS 3811 as it introduces the notion of value, which is linked with life expectancy, and requires consideration of the complex mechanisms which either erode or enhance the value of a built asset over time.

The usefulness of both definitions will depend on whether an economic/financial appraisal is being undertaken, or if the real concern is with operational and building condition issues.

The question also arises as to whether or not 'improvement' should be included under the heading of repair and maintenance. In any maintenance operation there will almost always be an element of improvement, if only through the replacement of an obsolete component. Thus, to a large extent, maintenance and improvement are inseparable, but in principle are clearly distinguishable from conversion, rehabilitation and refurbishment, which have the clear objectives of adapting or increasing the utility of a building, rather than maintaining it at the current level. This distinction between maintenance and improvement is of more than passing academic consequence as it may have a significant influence on funding arrangements. Housing Associations, for example, have until recently been able to attract grant aid for major improvements, whilst maintenance was funded from revenue.

The most sensible approach to take at the outset, is to see maintenance work as that which enables the building to continue to efficiently perform the functions for which it was designed. This may include some upgrading to raise the original standards, where appropriate, to contemporary norms and the rectification of design faults. Thus building maintenance needs to be seen as a part of a larger property management function and viewed in the context of the emerging discipline of facilities management (FM).

Both definitions of maintenance imply an interest in actions, which may be either physical or organisational. It is with the latter that this text will be primarily concerned and so it is necessary to consider maintenance within the micro-business environment of the company or organisation. Because building maintenance operates within a complex macro-climate, the level of activity will be determined by a host of inter-related factors. These can broadly be categorised into:

- Supply side factors, concerned with the ability of the industry to execute defined maintenance requirements
- Demand side factors, associated with the definition of maintenance tasks and the ability and willingness of building owners to execute them

Demand for construction work

The determination of demand for construction activity is a very complex affair and heavily dependent on the policies of government, either directly through its intervention in the public sector, or through its influence on the general level of economic activity in the private sector. To the economist, demand is regarded as the requirement for goods and services that the customer is able and willing to pay for. It is, thus, important to distinguish between demand and need[3].

In terms of maintenance, need can be measured by reference to the standard of building condition that is regarded as being acceptable, and the extent to which the actual condition falls below this standard. At any one time there may be a large quantity of latent demand awaiting the right conditions for its realisation. These conditions will more often than not be financially related.

Hillebrandt[4] identifies four major requirements for the creation of demand:

- ❏ The existence of a user, or potential user
- ❏ Someone willing to take on ownership responsibilities
- ❏ Available finance
- ❏ The existence of an initiator, such as a professional building manager

In addition to these factors being present, it is also necessary for contextual conditions to be favourable and thus external agencies, the most important of which is government, will be extremely influential in determining when they arise.

The supply of construction services

From time-to-time high levels of demand expose an inability of the industry to deliver the service required by its clients. This normally manifests itself in labour shortages and extended delivery dates. The reasons why this occurs are complex, but appear to stem from the cyclical nature of the construction economy, a lack of responsiveness to the cycles, and difficulty in attracting talented people into the industry. Such times are characterised by inflated tender figures and a tendency to seek less labour-intensive ways of building; sometimes with unhappy results, e.g. the system built housing schemes of the 1960s.

Demand for construction work is always extremely varied and the characteristic structure of the industry, with its diversity in size and type of company, has evolved in order to supply this diverse demand. When demand is low the industry attempts to adjust, and contractors will compete outside their normal markets. This may lead to involvement in work for which they are not necessarily suitable. Periods of recession lead to a slimming down of the work force and there is a tendency in construction for good quality labour to be lost permanently, which in turn tends to exacerbate inflationary problems in an upturn[5].

The structure of the construction industry

The construction industry has many characteristics which, taken on their own, are not unique in industry as a whole. The peculiarity of construction, compared with other industries, stems from the existence of a unique combination of these factors.

The industry is one of paradox in that, on average, its annual output contributes approaching half of the fixed capital formation of the country, and yet it is characterised by the presence of a very large number of small companies. When the size of the construction industry is measured in terms of its contribution to the Gross National Product (GNP) it can be seen that, even in a lean year, this is around 10%, which is roughly two and a half times the value of the output from agriculture, forestry and fishing.

Its product is large, heavy, and immovable, ranging from small-scale housing and extensions, through large sophisticated office and hospital buildings, to vast civil engineering projects. Most of its output is custom built and geographically widely distributed. In terms of production, its shop floor is a constantly changing one, and is characterised by the movement of staff, machines and materials to the emerging product, rather than by production-line processes. It therefore suffers more difficulty in mechanisation and rationalisation than most other industries.

The output of the industry is extremely variable, and the idea of construction being used as the regulator of the economy is well known[6]. Figure 1.1 shows output from 1978 to 1993, indicating large fluctuations. At the time of writing (1994/95) the industry has suffered further reversal due to economic demand side factors.

Another important measure is the number of people employed. This fluctuates either side of 1.5 million, accounting for 6–7% of the work force. The number of people employed evidently fluctuates with the industry's output. However, there is some evidence of a decline in real terms. Of particular concern for maintenance operations is an adequate supply of well trained labour.

No other industry is more fragmented or diverse in its structure, and it is still easier to form a construction company than almost any other. Figures are notoriously unreliable, but housing and construction statistics suggest there are in the order of 195 000 construction companies, of which approximately 70 000 are classed as general builders. However, of these firms, in the region of 80% employ fewer than ten persons whilst less than 1% employ more than 250 personnel. In 1993 there were in excess of 90 000 single employee companies and only 33 employing more than 1200[7] workers.

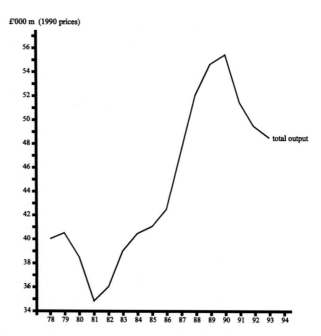

Figure 1.1 Construction industry output. (*Source:* Housing and Construction Statistics)

Building maintenance in the construction industry

Supply and demand

To create demand for maintenance activity there are a range of factors, which will be discussed later in this chapter. For the time being, however, it may be assumed that the general environment favourable for demand creation is as relevant to maintenance work as it is to construction activity in general, although the mechanisms at the micro-economic level may be different.

The influence of the following factors in particular should be considered:

❑ The general economic climate and its trends in respect of:
  ○ interest rates
  ○ inflation
  ○ industrial output
  ○ exchange rates
❑ The cost of construction in relation to the cost of other goods or services

❑ The operation of the rules and regulations governing land use and property development

Within the construction industry, the various sectors compete with each other for resources. This is true of maintenance activity, which, because of the way it is viewed by the various parties, may be at a relative disadvantage in this respect. Additionally, maintenance has to compete for resources in another market, that of the general business environment. Within both markets the supply of financial resources is clearly important, but within the construction industry itself there should also be some concern with regard to the supply of physical resources, such as an adequate and properly trained work force. This is a complex issue, and is dependent on the levels of available employment opportunity generally, and hence the level of demand for construction activity in particular.

Maintenance output

There is a belief that maintenance output remains fairly static within the overall peaks and troughs of construction output. To test the validity of this, reference can be made to the construction industry statistics, which separate out repair and maintenance work from other construction output. However, these statistics need to be treated with caution, as the work of maintenance departments is not always clearly identifiable due to the complexities of definition.

Figures from Building Maintenance Information Ltd (BMI) indicate that total expenditure on maintenance in the UK increased by 3.7% from 1988 to 1989 and by more than 33% in the ten years from 1979. From 1982 to 1989 the growth in maintenance spending outstripped the rate of general growth in the economy and reached in excess of 5% of GDP[8].

These figures were derived from Housing and Construction Statistics, and include an estimate for householders' expenditure on DIY materials taken from the UK National Accounts. Such estimates need to be treated with some caution, and the graphs and comments below are generally based on Housing and Construction Statistics alone.

Figures 1.2 and 1.3 show repair and maintenance output in comparison with the total output of the industry and there is a suggestion of less volatility. Between the 1950s and 1982 repair and maintenance, including housing improvement, increased its share of construction output from 25% to around 47%. Since then there has been a small decline but the figure remains well over 40%, (Figure 1.4).

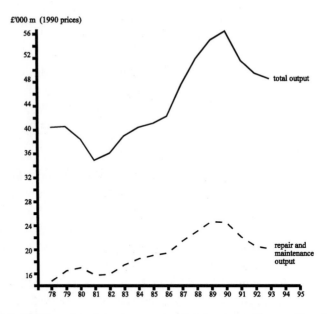

Figure 1.2 Repair and maintenance against total output. (*Source:* Housing and Construction Statistics)

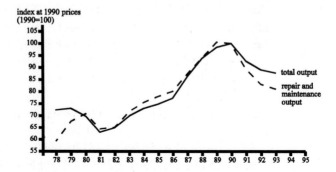

Figure 1.3 Repair and maintenance against total output indexed at 1990 prices. (*Source:* Housing and Construction Statistics)

These percentage figures on their own do not of course give the full story as, even at a constant level of maintenance activity, they will rise if overall output declines. For example, the fact that much of this increase occurred in a period of recession for the industry as a whole has prompted claims that the industry's resources are switched to this sector when new work is hard to come by. This claim is difficult to fully substantiate, not only for the reasons given above, but also because both supply and demand factors are influential.

It should be noted that, when the period under consideration began,

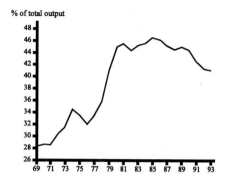

% of total output

Figure 1.4 Repair and maintenance as a percentage of total output. (*Source:* Housing and Construction Statistics)

there was still a substantial shortage of buildings as a result of World War II. By the mid-1970s this demand was slowing down, whilst at the same time the availability of land for new building was under pressure. More recently, there has been increased dissatisfaction by the general public with much new building, which has focused attention on the better use of the existing building stock. Thus, although a substantial proportion of the stock has become unfit for the use for which it was designed and built, this has created demand for adaptation and refurbishment. These pressures on the realisable demand for new work tend to increase the latent demand for maintenance work in order to extend the life of the asset, even though general economic conditions may not be favourable. Any changes in the proportion of maintenance work over this time should also be judged against a low base level, representing a very minimal level of maintenance activity at the beginning of the period.

More recent figures from BMI, based on National Economic Development Office (NEDO) forecasts[9], show that in 1991 maintenance output fell by more than new work output, and figure 1.3 shows this continuing into 1993. This fall appears to be distributed evenly across all sectors (Figure 1.5).

In the private sector, because the expenditure on existing buildings is likely to be internally financed, the user, owner, initiator and financier may be the same. The property industry may differ, but at this stage it is assumed for the sake of simplicity that free market pressures on rents will bring user and owner into harmony. The decision making process is therefore likely to be different to that in the public sector. Much maintenance work in the private non-residential sector is thought to be determined by profit levels, and at the time of writing these are under huge pressure[10], [11]. Indications of an emergence from recession during

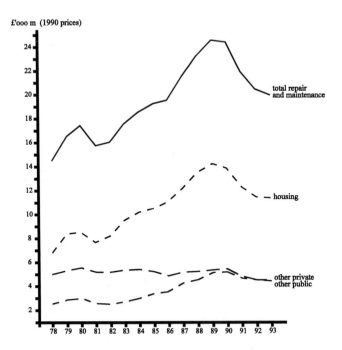

£'000 m (1990 prices)

Figure 1.5 Breakdown of repair and maintenance output. (*Source:* Housing and Construction Statistics)

the middle of 1993 confuse the issue somewhat, and forecasts tend to be somewhat erratic.

In the public sector, the level of maintenance activity among the various parts is worth examining, as each of these is the subject of strong political pressures from time-to-time, which can distort the distribution of maintenance activity. Housing is an obvious example, and the need for improvement in this sector is well documented in the English House Condition Surveys (EHCS) of 1981, 1986 and 1991, although there are doubts as to whether these produce a genuine picture of reality[12]. The publication of the 1981 EHCS report in particular, led to considerable pressure and increased activity in maintenance work on housing, which may have had the effect of increasing maintenance output, or led to a switch of finance from other areas, e.g. schools' maintenance. There is support for this view in figure 1.5, which shows a breakdown of repair and maintenance work for the period 1978 to 1993. Figure 1.6 also indicates that housing maintenance as a percentage of total repair and maintenance output increased steadily from 1982 to 1988, when it began to decline, in output and as a percentage. Figures for 1992 to 1993 show a small increase in percentage terms, but this is against a backdrop of declining total repair and maintenance output.

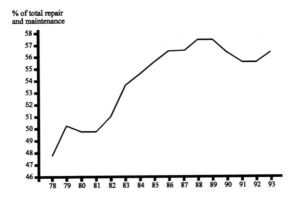

Figure 1.6 Housing as a percentage of repair and maintenance. (*Source:* Housing and Construction Statistics)

When examining maintenance statistics it should be noted that the figures given by BMI include a contribution for maintenance work carried out by directly employed labour in private sector organisations, and for DIY housing maintenance, neither of which are included in the Housing and Construction Statistics. Estimates of these contributions are based on UK National Accounts[13].

A comparison of maintenance expenditure in relation to the replacement value of the building stock, using data from BMI, in figure 1.7, reveals some erratic trends, but a clear decline with respect to housing maintenance. Housing maintenance expenditure, in relation to the value of the stock, has been consistently higher than for either public or private non-housing, although the figures do include some alteration and improvement work. If the estimated value of DIY labour is included, the percentage rises by around half a percentage point. Whilst it must be borne in mind that stock valuations are rather imprecise the trend

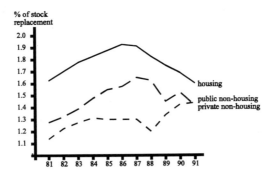

Figure 1.7 Maintenance as a percentage of stock replacement. (*Source:* Housing and Construction Statistics)

appears to be clear. The raw statistics also say little about the condition of the building stock, for example its age profile, and there is a strong case for arguing that maintenance expenditure, expressed as a percentage of stock value, should be increasing when this is taken into account[14].

Government policy and maintenance

Since 1979 the role of government, in terms of economic management, has moved away from Keynesian principles, with its emphasis on demand management, and returned to the classical theories espoused by Adam Smith and others. Government thus adopts a far less interventionist approach and relies on supply and demand controlling the market, with government using indirect means of influence, such as interest rates, money supply and fiscal measures. It is useful, therefore, to consider the influence of some of these factors on the level of maintenance activity.

Taxation

This is of some significance, since maintenance expenditure by businesses is classed as revenue expenditure and is allowable against income and corporation tax, whereas capital expenditure is not chargeable on commercial buildings. It has been argued that this does not encourage an increase in capital expenditure to reduce maintenance, as the real cost of subsequent maintenance is thereby reduced. Conversely, it can also be argued that if a firm saves on capital expenditure, it may as a result increase its tax liability, either through interest earned or foregone by reduced borrowing. Movements in the level of taxation, however, will influence expenditure on buildings already in use; and there is a view that at higher levels of taxation increasing revenue expenditure to reduce tax liability is attractive[15].

However, the savings that can be made in this respect are heavily dependent on profit levels against which outgoings can be set. At the time of writing these are under heavy pressure and thus taxation policy is likely to have only a minimal effect.

Interest rates and inflation

The control of inflation has been the cornerstone of the government's economic policy for some time and interest rates one of the prime measures of control. Inflation is measured by the Retail Price Index

(RPI) and a monetarist policy to this implies the use of interest rates as the controlling mechanism. The theory is that in times of high inflation high interest rates are used to bring it down. If this succeeds then lower interest rates may be used to stimulate the economy.

Following this economic theory it might be assumed, therefore, that low interest rates will encourage maintenance expenditure. However, as figure 1.8 shows, this is by no means certain. One reason for this is that maintenance expenditure only yields an indirect return that is not readily represented by cash gain, except, perhaps, if the property is to be sold. In the main the return is a long term one.

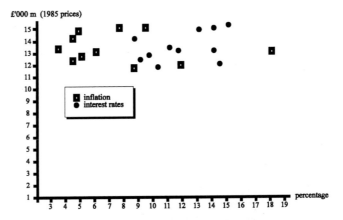

Figure 1.8 Maintenance expenditure versus inflation and interest rates.

Capital projects are far more responsive to changes in interest rates because of the effects on financing costs. High interest rates encourage the reduction of capital expenditure, perhaps at the expense of increased expenditure in the longer run. It is difficult to track this cause and effect cycle and thus this hypothesis has not been completely proven.

In terms of its relationship to inflation, figure 1.8 indicates that maintenance expenditure exhibits little correlation[16].

Economic confidence

Future expectations within the economy may affect maintenance expenditure and industry looks to government, and other predictive indicators, for a lead in this respect. The building owner will thus set future investment levels and budgets in this context and, certainly within the private sector, this will affect maintenance budgets and hence real expenditure. There is evidence in BMI analyses that this is a factor at the time of writing[17].

Correlation with economic variables

A correlation analysis (Appendix 1), relating total maintenance expenditure with a number of economic variables, has been carried out[18] and the following results obtained:

Gross Domestic Product	0.999
Construction output	0.993
Average interest rates	0.024
Mortgage rate	0.138
Average RPI	−0.520

This indicates a strong statistical relationship between maintenance expenditure, GDP and construction output, but little with interest rates and RPI.

Maintenance needs

Demand factors generally

Demand for maintenance output is subject, with some minor differences, to the influences identified above, and there is undoubtedly a significant latent demand awaiting the right conditions for its realisation. In the first instance, it can be generally recognised that there exists a need for maintenance work to put and keep the building stock in an acceptable condition. The precise extent of this requirement is only measurable in general terms. There are two reasons for this:

❑ A lack of universal agreement as to what represents an acceptable condition
❑ The absence of a fully comprehensive picture of the present condition of all our buildings, although reports focusing on particular building types are published from time-to-time, especially for the public sector

At any one time some of this need for maintenance work is being satisfied. How much work is carried out reflects the willingness of clients to buy this service and their ability to do so. It is important to distinguish between the ability to pay and the willingness to do so. To create maintenance activity both must exist. There may be a perceived lack of maintenance work, due to an inability to pay resulting from unfavourable economic or business conditions, or due to a building owner being unwilling to buy, even when finance is available.

A strong body of opinion exists that considers there to be a general lack of concern to repair and maintain buildings to an acceptable condition. This under-resourcing is presumed to be the result of a massive level of ignorance and/or complacency. These rather general observations are supported by a number of reports, notably by the Audit Commission, which focus on the maintenance of the building stock in various parts of the public sector.

Housing

Housing and Construction Statistics indicate that housing maintenance represents over 40% of repair and maintenance expenditure. This figure excludes a substantial input from the owner occupier sector, especially that executed on a DIY basis.

The housing stock can be divided into four categories:

- Local authority
- Housing association
- Private rented
- Owner occupied

There is other housing, notably that associated with the armed forces, for which detailed data is not readily available. An interesting point, worthy of note here, is that in Scotland local authority housing represents around 44% of the total housing stock, compared with 22% in England and 2–3% elsewhere in Western Europe.

Whilst the National House Condition Surveys and the Audit Commission reports on local authority housing provide valuable data on the condition of some of the housing stock, there is a general lack of reliable data on owner occupied properties, and those in the private rented sector. This makes informed comment on housing in those areas rather difficult, and only generalisations can be attempted.

Up to 15 million people live in what may be termed public sector or social housing in England and Wales, and the quality of their lives is directly affected by the condition of their home and the environment of the estates in which they are located. Housing represents not only an important social and political issue, but also an economic one, when one considers that in 1986 the replacement cost of the stock was calculated to be £100 billion.

The National House Condition Surveys of 1981[19] and 1986[20] presented a very bleak picture of the condition of the stock. The details of these reports were widely publicised and there is a suggestion, statistically, that the earlier report did have some effect in raising the profile of housing maintenance within the construction industry, and may have been partially instrumental in encouraging a more professional attitude towards maintenance management. This is by no means to argue that all is well, but there has certainly been some progress, as is borne out by Royal Institution of Chartered Surveyors (RICS) figures indicating the increasing level of involvement by Chartered Building Surveyors[21].

In 1986 the Audit Commission published two reports, *Improving Council House Maintenance*[22] and *Managing the Crisis in Council Housing*[23], in which they described the scale of the problems facing many housing authorities, particularly those in urban areas, and notably Inner London. The crucial element of the problem was identified as the backlog of maintenance, repair and improvement work necessary to bring dwellings up to an acceptable standard. This backlog was estimated, in reports by the Department of the Environment (DoE), to be £20 billion in England alone, despite an average annual expenditure of about £425 per dwelling.

There were notable variations identified in the report. For example, in Inner London the maintenance backlog averaged £4500 per dwelling compared with £1400 in some shire districts. These differences were due to a variety of factors, one of which was relative age profiles, and another the extent to which backlogs were a function of previous underspending.

For non-traditional housing, the bulk of which could not be said to be old in housing terms, the average backlog was about 70% higher than for traditional construction. A large part of the problem was associated with the system built dwellings of the 1960s and early 1970s, the majority of which were concentrated in urban areas.

The reports pointed out that steps were necessary to deal with the likelihood of the backlog growing, and they estimated that, without immediate action, new problems that would continually arise on the newer portion of the housing stock would add of the order of £900 million to the maintenance bill in the next 15 years.

Whilst the general thrust of these Audit Commission reports was directed towards the need to improve management practice, so as to make more efficient use of resources, they acknowledged that additional sources of funding would be necessary. The major recommendation was that local authorities should draw up a five year plan, and fund the programme at local level through rents and revenue from the sale of assets. A further recommendation was that the existing systems for controlling revenue and capital should be changed, limiting borrowing to the financing of programmed repairs and leaving rent income to cover day-to-day maintenance. It was concluded that improved management could generate an estimated 30% in improved maintenance value.

Another view of the problem was presented in a report by the Association of District Councils (ADC)[24], which set out with the main objective of trying to assess progress in the interval between the EHCS of 1981 and 1986. One of the major comments was a strong criticism of the DoE for using new methodology in the latter report, particularly with respect to the definition of disrepair, which excluded moderate disrepair, and thus had a tendency to understate the extent of the problem. It also argued that the findings were not based on a large enough sample.

The 1986 survey used the following main indicators to describe stock condition.

(1) Unfitness

This was defined by s.604 of the Housing Act 1985 as a dwelling deemed to be so far defective in one or more of the following matters as not to be reasonably suitable for occupation:

- ❑ Repair
- ❑ Stability
- ❑ Freedom from damp
- ❑ Natural lighting
- ❑ Ventilation
- ❑ Water supply
- ❑ Drainage and sanitary conveniences
- ❑ Facilities for the preparation and cooking of food
- ❑ Disposal of water
- ❑ Internal arrangement

(2) Lack of basic amenities

This was defined as the absence of one or more of the following:

- ❑ A kitchen sink
- ❑ A bath or shower in a bathroom; a wash hand basin
- ❑ Hot and cold water provided for each of these and an indoor WC

(3) Poor repair

This was defined as where urgent repairs to the external fabric of the dwelling were estimated to cost more than £1000.

(4) Poor condition

This was a generic term used to describe a property which fell within one or more of the above definitions.

The 1981 survey employed a different definition of repair and also used an indicator called serious disrepair, defined as a dwelling requiring over £7000 worth of work (at 1981 prices). To aid comparison, the DoE carried out a smaller separate 1986 survey of the dwellings surveyed in 1981, using the same methodology as the 1981 survey. This indicated that the number of houses lacking basic amenities reduced from 5% to 2.9%, but showed only a small decrease, from 6.3% to 5.6%, in the number of unfit properties, and from 6.5% to 5.9% for those in serious disrepair.

This survey also showed that in the 1981–86 period only 42 000 houses classed as unfit were cleared or closed, out of a total number of 900 000 in 1986.

The ADC[25] commented that:

'At the present rate of clearance it would take over 100 years to replace the number of unfit dwellings . . .'

They noted that the absolute number of unsatisfactory dwellings (those either unfit, lacking basic amenities or in poor repair), at 2.9 million or 15% of the stock, represented cause for grave concern. This figure was also at variance with the official figure of 910 000 unfit properties given in the EHCS, which highlights the great difficulty involved in assessing real needs.

The 1986 survey showed that pre-1919 dwellings were in the worst condition, and that more than half of these were in the owner occupied sector. The ADC, however, also expressed serious concern about the increasing deterioration of many inter-war years houses. In general, the survey found that poor housing continued to be occupied by low income families. The proportion of poor housing was higher in rural areas (22%) than urban areas (14%) and that, regionally, the North, Midlands and South-West fared worst.

Local authorities owned over half of all post-war dwellings in poor condition, and the DoE survey concluded that this was a result of poor initial quality and neglect in terms of planned maintenance. It also revealed that many authorities did not have accurate records of the condition of their properties, which is essential if proper programmed maintenance is to be carried out.

On the question of funding, the survey indicated that improvement grants were reasonably well targeted, but the ADC were concerned that 25% of the dwellings which had been the subject of individual or area action in the years 1981–1986 had fallen back into poor condition by the end of the period. This may be explained not only by the standard of workmanship, but also by a failure to see maintenance as an on-going rather than a one-off commitment.

With respect to resourcing the ADC said that they:

'... find it incomprehensible that unlike the earlier 1981 survey and the 1986 Welsh survey, the 1986 English survey does not estimate total repair costs.'

They went on to estimate themselves that:

'... on the basis that there has been no significant change in the level of disrepair over the survey period it can be calculated on the basis of the Government's 1981 figures that at today's prices the total repair bill in England stands at around £50 billion.'

The RICS have also commented[26] that:

'For too long the housing stock – both public and private – has suffered from inadequate funding for maintenance.'

The DoE, in a 1985 enquiry, indicated that there was a need for a sum of £22 billion (1989 prices) for the repair and renovation of local authority owned stock in England, and at current levels of provision this would take 11 years to complete. These figures, however, failed to take into account additional dwellings falling into disrepair in the meantime.

Private landlords make a relatively small contribution to the repair and maintenance of their dwellings, and the average cost of outstanding repair work is twice that of any other tenure. Another particularly hard hit sector comprises houses in multiple occupation, and here the problem is compounded by inadequate or inconsistent data on the number of families in such accommodation.

The house condition surveys take account of all housing, but there is less substantive data available for owner occupied stock. The DoE survey pointed out that the condition of owner occupied dwellings gave cause for concern, and that owner occupiers do not invest sufficiently in maintaining the condition of their property. There is some evidence suggesting that, in this sector, there is a substantial level of neglect, probably largely due to ignorance. In recent years there has been a steady increase in the proportion of owner occupiers, particularly amongst lower income families, and concern has been expressed at this, as many of the dwellings in this group may have been in questionable condition at the time of purchase.

The ADC recommended a government led campaign, involving the building industry and local authorities, that emphasises the need for timely and effective maintenance and repair.

The 1991 EHCS[27] only served to add further confusion, in that another set of criteria, based on the requirements of the 1989 Local Government and Housing Act, was used to determine unfitness. Whilst the 1991 unfitness criteria had the same basis as the previous standards, they incorporated a number of important differences. In the 1986 survey, the unfitness requirement of internal arrangement had always been seen as unsatisfactory. The later standards require a dwelling to:

- Be structurally sound
- Be free from serious disrepair
- Be free from damp prejudicial to occupant's health
- Have adequate lighting, heating and ventilation
- Have an adequate supply of wholesome water
- Have suitable food preparation facilities and a sink with hot and cold water
- Have for the exclusive use of the occupants a suitably located bath or shower and wash hand basin, each with a supply of hot and cold water

A failure to meet any one of these criteria renders a dwelling unfit. Using these criteria, the 1991 survey found that there were 1.46 million unfit dwellings. By constructing a multiple regression model, to permit an indirect comparison between 1986 and 1991, the report claims that if the new standard had been applied in 1986, the number of unfit dwellings would have been 1.64 million. There is no way of substantiating the validity of the regression model used, and some caution needs to be exercised when considering some of the conclusions drawn from it. These can be summarised as follows:

- There had been a substantial improvement in the condition of the pre-1919 stock, but the majority of unfit dwellings still fell within this group
- Most of the improvements in condition were in areas which in 1986 had the highest levels of unfitness
- Unfit dwellings occurred in new stock, but the problem was largely still in the public sector
- The most common reasons for unfitness were disrepair, and inadequate facilities for food preparation and cooking
- Half of the unfit dwellings were unfit in one requirement, or had a low cost of returning them to fitness
- The average cost of the minimum works to make a dwelling fit was given as £2100, but there was a very large spread

The 1991 report has been criticised severely because of its failure to facilitate easy comparison between reports, and also because of the absence of a definitive estimate of the maintenance backlog in monetary terms.

A later report by the Audit Commission[28] estimated the maintenance backlog on the four million council properties in England and Wales to be £8.5 billion, compared to £13.5 billion in 1985, but expected the situation to deteriorate due to constraints on local authority spending.

Health care buildings

Health care buildings represent, perhaps, the most difficult group of largely public sector buildings to maintain because of their complex engineering services and their heterogeneous nature. Furthermore, safety and hygiene considerations make the condition of these buildings a particularly sensitive issue.

At the time of writing, there is much uncertainty amongst buildings and estates managers due to the move towards the formation of self-governing trusts, which further complicates already complex funding mechanisms. Part of the development towards trust status necessitates a great deal of survey activity to create asset registers. So far the bulk of this activity has been targeted at mechanical services and equipment, as this represents the larger portion of the asset base, and is the most critical in terms of patient care. Nationally, no substantive comprehensive survey data is available on the condition of health care buildings. However, in 1986 the total backlog of repair and maintenance in hospitals was estimated to be in the region of £2 billion[29].

An Audit Commission report in 1991[30] estimated the backlog at £1 billion, and was very critical of maintenance practice in the National Health Service (NHS). Amongst their conclusions they stated that:

❏ Some planned maintenance programmes were out of control and that there was evidence of too great an emphasis on preventive maintenance
❏ There was too little competition from private contractors allowed for in allocating maintenance work
❏ Too great an emphasis was placed on new build at the expense of refurbishment options

Amongst maintenance professionals there is a strong feeling that maintenance is underfunded, and that little preventive fabric maintenance is taking place. For example, one maintenance manager estimated that the large hospital for which he was responsible had a £4 million maintenance backlog needing rectification over a five year programme. For this period he had been allocated approximately 25% of this sum, with the bulk of the funding coming at the end of the period, by which time he expected that the maintenance backlog, due to continuing neglect, could increase by as much as 50%. Virtually all of his financial allocation for maintenance was used up by planned preventive maintenance on engineering services, much of it driven by the need to meet statutory and insurance requirements. If this scenario was repeated on a national scale then there would clearly be a major problem in the making. It is difficult at the time of writing to judge whether the move towards self-governing status will alleviate or exacerbate the situation.

Educational buildings

There are approximately 25 000 primary and secondary schools in England and Wales managed by local authorities. Within a statutory framework, local authorities have operational responsibility for school building maintenance, and also have discretion, at local level, for education expenditure and the determination of priorities. They must also, however, determine their overall expenditure within the spending guidelines imposed by central government, which may also indicate the priority it attaches to education through that portion of the annual local authority spending settlement.

There are numerous publications of a non-statutory nature from central government that, as well as providing advice and guidance for local authorities, give an indication of central government concerns. The bulk of central government's contribution to local authority spending is through the Revenue Support Grant, which is not allocated to specific services. Building maintenance funding for all buildings, including schools, must therefore compete for its allocation with other services, many of which may be perceived as having a higher priority, both politically and socially. Maintenance expenditure is thus governed by the supply of funds, rather than on the basis of building need.

Condition surveys carried out by many local authorities on their education buildings have the primary objective of providing a rational basis upon which to divide up their budgets and, whilst comprehensive in terms of coverage, they are superficial in detail. There is growing unease over the condition of this large and diverse group of public sector buildings, and this was the subject of a National Audit Office report published in 1991[31].

Under the Education (School Premises) Act of 1981, local authorities were required to bring all existing schools to a defined acceptable standard by 1991. This period has been extended to 1996, and a review of the regulations will be undertaken in the context of the changes that have taken place in educational practice and management.

In 1985 the Department for Education and Science (DES) (now the Department for Education and Employment) provided a paper for the

National Economic Development Office (NEDO)[32] in which they noted that the state of maintenance in educational buildings varied widely from one local authority to another. Moreover, it stated that a significant number of pupils were being taught in schools that were in an unsatisfactory condition, and pointed out the detrimental effect this was likely to have on achieving educational standards. The standards of management used in the control and execution of maintenance work came in for criticism, and the paper indicated that financial resources were not being used in an optimal manner.

In 1986 the DES undertook a survey of 800 schools[33] and concluded the following:

- 38% of the primary schools sampled required work on roofs and heating systems
- 32% reported deficiencies in walls and windows
- 24% of secondary schools had broadly similar problems
- In all areas a backlog of repair had built up and it was estimated that the cost of bringing school buildings up to the standards required by the 1981 Act by 1991, adjusted for predicted pupil numbers, was £2 billion at 1987 prices
- The cost of structural repairs and maintenance alone was estimated to be between £730 and £995 million

In 1988 and 1990 the National Foundation for Educational Research was commissioned by the National Audit office to carry out research into the problem[34]. They concluded in their 1990 study of 12 local authorities that:

'... the difference between what authorities are spending and what they would like to spend on maintenance work remains substantial. Overall, the findings of the survey commissioned by the National Audit Office underscored concern by the local authorities about the condition of school buildings and the backlog of repairs as originally identified by the survey work undertaken by the Department.'

Housing and construction statistics do not separately identify repair and maintenance expenditure on educational buildings, but the Audit Office highlighted the very heavy cost burden. Their figures showed that in 1988/9 £340 million from revenue was spent on repair and maintenance in primary and secondary schools and an estimated extra £40 million by governors in voluntary aided schools. Additionally, a substantial portion of a £2 billion capital expenditure programme in the period 1986/7 to 1988/9 would have been devoted to the remedying of defects and to general improvements. If these improvements are taken as being required to bring building conditions to the standards required under the 1981 Act, then a substantial portion can be counted as repair and maintenance.

Research and discussions with local authorities revealed that priority was given to safety, security and essential repairs, and that routine maintenance and decoration received less attention. This undoubtedly leads to increasing rates of deterioration and accelerates the growth of maintenance backlogs. 1984 DES guidelines suggested that the balance between planned and emergency maintenance should be in the order of 65:35, whereas at the time the report was prepared the balance was approximately 55:45.

The Audit Office report also referred to the subject of estate management and earlier advice from the Audit Commission on this topic[35]. In the light of demographic changes, the need for Local Education Authorities (LEAs) to adopt a professional management attitude was emphasised, so that the rationalisation and provision of maintenance expenditure could be more focused.

Another issue identified by the Audit Office related to the design, type and age of stock. The report clearly identified major problems associated with post-war prefabricated and system built schools, and it is claimed that the need to divert funds to the execution of urgent maintenance work on these has starved many older, but basically sounder, schools of maintenance funds. The largest building programmes were in the period 1970–74, and many items, such as boilers and wiring, are at the end of their design life. This may lead to a peaking of maintenance demand in the mid-1990s.

In addition to the difficulties outlined above, the report also presented startling evidence on the contribution made to maintenance workloads by the growing incidence of vandalism. There is a suggestion in the report, based on a survey of 17 local authorities in 1988, that vandalism and arson could be costing in the order of £100 million per annum at 1987 prices.

Informed comment on the report has claimed that the rate of deterioration of the nation's schools is much greater than it suggests. A report by the Association of County Councils[36] estimated that a capital investment of £1400 million was needed in Britain's school buildings, compared with official estimates of £500 million.

Many matters are likely to be brought to a head by health and safety considerations.

'... more than 80 schools in Strathclyde, Britain's largest education authority, were set to close by Christmas on health and safety grounds.'

The same article[37] quoted the controller of building services at Barking and Dagenham as saying:

'... schools are deteriorating faster than we think. What is happening at the moment is a lot of fire fighting repair work. If this continues, sooner or later schools will close because of a lack of repairs.'

An important contemporary issue is the change which has taken place in the management of schools, and in particular the delegation of responsibility for minor maintenance work to the governing bodies of individual schools[38].

Whilst the current debate is focused on school buildings, it is important to emphasise that these only represent one portion of the building stock of education buildings. The polytechnics and larger higher education institutions became incorporated in 1990 to form Higher Education Corporations (HECs), and in 1992 the majority of them became new universities. Hitherto, these institutions had been under local authority control, and the cost burden of building maintenance fell on LEAs.

Incorporation effectively took them into the private sector, and whilst, on paper, the gifting of the buildings appears to be attractive, the new HECs have inherited considerable liabilities. Many of the polytechnics, for example, grew out of the development and merger of older inner city educational institutions which were accommodated in ageing buildings. Furthermore, over their lives, these buildings often suffered from a lack of attention being paid to the proper maintenance of their fabric. Additionally, the current expansion in the participation rates for students in higher education is imposing severe pressure on space in these buildings, leading to very intensive patterns of use.

The traditional universities, which have operated on a more independent basis for many years, have developed more effective estate management regimes but, notwithstanding this, there is growing concern about the ability of all universities to meet increasing maintenance needs as their building stock ages.

Execution of maintenance work

The agencies which carry out repair and maintenance work are contractors and, in both private and public sectors, direct labour organisations (DLOs). Figure 1.9 shows the breakdown of the execution of maintenance work between contractors and DLOs. The proportion of work executed by contractors has increased steadily from around 70% at

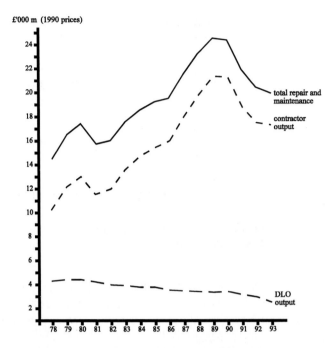

Figure 1.9 Execution of repair and maintenance. (*Source:* Housing and Construction Statistics)

the beginning of the period to over 80% in 1988, since when it has remained relatively steady.

Figure 1.10 shows the breakdown of DLO output over the same period. Total output has declined in real terms, with a steadily decreasing repair and maintenance output. Figure 1.11 reveals that, within this overall scenario, housing maintenance is relatively stable, and that the overall fall in output is caused by reductions in work to other public

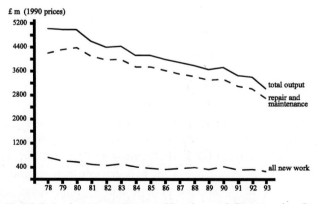

Figure 1.10 DLO work output. (*Source:* Housing and Construction Statistics)

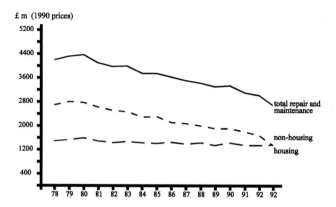

Figure 1.11 DLO repair and maintenance work output. (*Source:* Housing and Construction Statistics)

sector properties, such as educational and health care buildings.

The statistics available leave many unanswered questions, but figure 1.12 does indicate increasing activity of private contractors in repair and maintenance work.

Although a significant amount of construction is also undertaken by a thriving DIY sector, its contribution to output is not included in government statistics, even though it is considered to be a major contributor to repair and maintenance output.

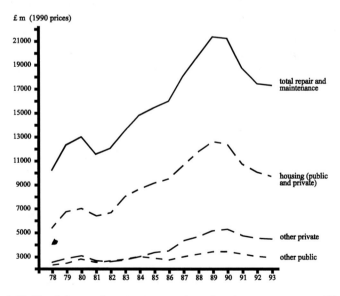

Figure 1.12 Breakdown of contractor repair and maintenance output. (*Source:* Housing and Construction Statistics)

Building maintenance and the professions

A major characteristic of the construction industry is the multiplicity of parties who may be involved in any construction project. There are several reasons for this, but the major contributing factor is the traditional division between design and production, which typifies the industry and leads to a separation of the design-related professions from those more closely associated with production.

'In no other industry is the responsibility for design so far removed from the responsibility for production.'[39]

Whilst it is normal to see the architectural and the contracting professions as epitomising the conventional divisions, there are further divisions within these groupings. Even allowing for contemporary thinking about procurement, which attempts to address traditional divisions, it would be wise to accept that there are differences between architects and builders which are educational, social and cultural, and that these will not disappear overnight.

Another cause of fragmentation, in respect of responsibility for construction, is the range of clients served by the industry. Large client organisations may also employ professional advisors, in addition to the architect, to assist in managing complex projects. In some cases the building owner and the building user, who are not necessarily the same, may both be involved in the procurement process. When the position of building maintenance within this tangled web is considered a complex picture results.

The assessment of repair needs and the execution of maintenance of a building in use, brings into play another set of parties who are rarely related to the team who procured and constructed the building, perhaps with the exception of the building owner, who may or may not have had some influence at the design stage.

The role of the building owner with respect to maintenance considerations during design will vary depending on their interest in the use of the building. The assessment and management of maintenance increasingly involves the chartered building surveyor at all stages of the process, and the Royal Institution of Chartered Surveyors (RICS) Building Surveyor's Work Load Survey of 1989/90 (figure 1.13) demonstrates this group's increasing involvement in repair and maintenance, in both the public and private sectors.

With the industry suffering the effects of severe recession at the time of writing, many private building surveying practices consider maintenance-related activities to be the key to their survival. The execution of

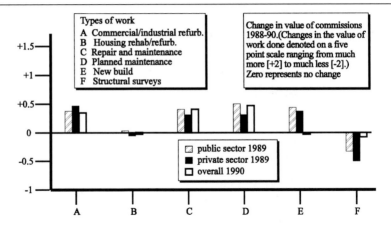

Figure 1.13 RICS: change in value of commissions. (*Source:* RICS workload statistics survey)

maintenance work itself, which usually makes use of a production team having no connection with the original construction team, is subject to huge variations on an industry-wide basis. The management systems employed, and hence the involvement of the professions, are hardly ever the same from one organisation to another.

Education and training for building maintenance

The report of the DoE Committee on Building Maintenance in 1972[40], contained a forthright section concerned with education and training needs. This report was 'revisited' in 1988, and formed the subject of a BMI special report[41].

The 1972 report identified the three following levels of education for maintenance:

- The education of all those who contribute at a professional and managerial level to building and maintenance processes
- Education for supervisory personnel who are in direct control of maintenance operations
- Education and training for those who undertake maintenance work

Unpublished work by the Polytechnic of Central London concluded that, whilst maintenance was not central to any of the objectives of the existing professional institutions, some indicated an increased awareness of its importance. There was, in the majority of educational programmes for built environment professionals, disappointingly poor coverage of such important topics as the design/maintenance relationship, economic evaluation and maintenance management. The study identified three main roles and six sub-roles, for which adequate education and/or training was needed. These are summarised below.

- ❑ Maintenance specialist
  - ○ building fabric
  - ○ building services
- ❑ General construction practitioner
  - ○ economics
  - ○ design
  - ○ inspection
  - ○ construction
- ❑ Clients

This represents the range of parties and functions associated directly, or indirectly, with the maintenance process, and was used as the basis for the design of an education programme in building maintenance, which was in turn influential on later provision by the Chartered Institute of Building (CIOB).

In terms of the design professions, there has been little movement by the Royal Institute of British Architects (RIBA), and there remains grave cause for concern with regard to the industry's performance in terms of the design/maintenance relationship.

The growth of the Building Surveyors Division of the RICS has helped to give increased attention to building performance, although this is often submerged beneath more glamorous activities such as project and facilities management.

Professor Tim Clark[42] comments on what he sees as a lack of bonding between people as the cause of many of the industry's ills. He goes on to say that:

'The key to the building up of an appropriate and regularly used network of training around the maintenance task is to obtain the commitment of its users through involving them in its planning. Planning promotes a sense of ownership and with it comes a fuller awareness of group identity, choices and opportunities.'

General experience in the operation of maintenance work, and the way in which it is included in courses of professional study, in many ways epitomises the ideologically fragmented nature of the construction industry. In considering the performance of the building stock, nowhere is it more evident that the industry lacks a basic sense of cohesion amongst its members. That a more integrative approach to education is desirable is just as valid for maintenance management as it is for building design or production in general.

For some years now there has been a general decline in the level of training within the industry as a whole, especially in craft training, which many people would see as having been diluted. However, this may be less of a problem for general maintenance work, where often the requirement may be for a multi-skilled operative, than in the more specialised areas of conservation work.

The real problem for the industry as a whole is one of volume of training and the future skill shortages this will create. Given its unfavourable position, in terms of competitiveness, when competing for scarce resources, maintenance operations may become adversely affected through lack of suitable manpower. The industry has been generally short sighted in its approach to training, not only in terms of output, but also in content. There must be a major shift in attitudes, and the impetus for this must come from the highest level.

In terms of improving performance, the major stimulant should in theory come from the client, and proponents of free market pressures would argue that, in the long run, the organisation which is most efficiently organised, with the best trained work force will prosper. If this were true one would expect to see substantially more effort being diverted to training than is at present evident.

Even if this set of problems can be resolved, there remain other issues specific to the training of maintenance operatives to be dealt with. It has to be recognised at the outset that the majority of maintenance operatives gravitate into maintenance work, after having spent a number of years in new build work, so that their experiential base is a general one. It has been acknowledged for some time that the requirements for maintenance operations are best met by multi-skilled operatives, and present training programmes do not produce this person. There is, in other words, little or no specific training for maintenance operatives. Even if such programmes existed, it is extremely unlikely that an industry which struggles to attract talented people for its more glamorous aspects, would succeed in doing so for what is often seen as its least attractive. The best that can be expected is that existing single-skill training programmes pay much more attention to the inter-relationship of skills, and attempt to inculcate a greater awareness in trainees of all aspects of the building process.

This inevitably places a quality premium on good supervision, which represents the point at which the professions and the operatives interface, and where many of the deficiencies of these respective groups have to be remedied. It is at this level where, perhaps, the greatest headway can be made, and where training can act to improve not only the skills of a group of people who are amongst the most dedicated in the industry, but also to raise their status by giving them much more explicit recognition.

Maintenance cost trends

Maintenance work is usually more expensive than new work due to the following factors:

❑ It is usually carried out on a small scale leading to diseconomies of scale
❑ The need to strip out existing work and generally prepare for repairs and replacements
❑ It frequently has to be carried out in confined or occupied places
❑ It is very common for the cost of accessing a maintenance item to be several times that of actually carrying out the repair
❑ The cost of making good and general clearing away is disproportionately high
❑ It incurs substantial disturbance costs on the operation of the building and perhaps lost production

The best co-ordinated source for maintenance cost trends is provided by Building Maintenance Information Ltd (BMI), a company of the RICS, which has replaced the earlier Building Maintenance Cost Information Service (BMCIS). To give a sensible picture, over a suitable period of time, requires the use of data from both these sources. Continuity and consistency between the two seems to have been satisfactorily achieved, although some revisions have been made in the method of computing the indices which are essentially the most appropriate method of representing cost trends.

Both BMI, and before this BMCIS, publish occupancy cost analyses of buildings, based on cost data provided by a range of building owners, in both the private and public sectors. There has consistently been a shortage of data to compile a sufficiently large sample of such studies, and those owners who have contributed, represent an unrepresentative sample in terms of organisational characteristics.

Furthermore, the data available does not represent a particularly broad range of building types, either in terms of occupancy characteristics, or type of construction. These statistical weaknesses have to be recognised if attempting to use data for cost predictions, and the distortions they can potentially cause need careful consideration in the production of building cost indices.

The methodology used in the production of the occupancy cost indices is described in BMI Special Report 190[43], and they are published regularly in the BMI Quarterly Cost Briefing. The indices are derived from published data on construction labour and material costs, and their production requires the determination of appropriate weightings for these inputs as outlined below. Special Report 190 describes how this has been carried out, and longer term maintenance trends are evaluated in regular special reports. At the beginning of 1990 the indices were re-based and cover maintenance, cleaning and energy costs.

Maintenance costs were analysed by BMI, using data from their own occupancy cost studies, to show the breakdown between redecoration, fabric maintenance and services. For all maintenance this revealed the following distribution of costs:

Redecoration	32%
Fabric maintenance	30%
Services	38%

To establish a breakdown of input costs within each of these elements BMI carried out further work, extending beyond data available from their own occupancy analyses, to give figures for the labour/materials split as set out below:

Fabric maintenance	65:35
Decoration	89:11
Services	65:35

For cleaning, BMI accepts the extreme variability to be expected in practice, and includes for plant in their breakdown. A breakdown between labour, materials and plant of 90:6:4 is used.

Having established these breakdowns, they are then applied to published cost indices. Special Report 190 lists in detail the cost indices that are used to produce maintenance cost indices using the above weightings. BMI also carried out a sensitivity analysis, to show the effect on trend interpretation of variations in these weightings, and concluded that, over the nine year period to 1989, the overall maintenance cost index would only vary by + or −0.5% either side of the 65:35 figure for fabric maintenance in any year if a split ranging from 50:50 to 80:20 was used.

In addition, they also tested the responsiveness of the indices to changes in individual labour and material cost inputs and concluded that, whilst material cost variations had only a minor effect, labour cost changes were much more significant, and for that reason strongly advise caution on the selection of appropriate labour cost indices.

Special Report 197 gives Maintenance Cost Indices for 1970 to 1990, using the first quarter of 1990 as a base. In compiling the basis for the indices, BMI produced a total occupancy expenditure study for four building types:

- Administrative and commercial facilities
- Recreational facilities
- Educational, scientific information facilities
- Halls of Residence

This was executed under the following cost headings:

- Decoration
- Fabric
- Services
- Cleaning
- Utilities
- Administrative costs
- Overheads

An average of all four types gave the following percentages:

Utilities	28.0%
Overheads	21.5%
Administrative	17.0%
Cleaning	16.5%
Fabric	6.5%
Services	6.0%
Decorations	4.5%

In general, utilities (fuel and energy) represent the largest proportion of annual expenditure and fabric, services and cleaning combine to amount to only 17%. However, these figures, based on a small sample, are distorted, in that many of these maintenance costs are derived from operations carried out on a cyclical basis, and are not therefore incurred every year. A more detailed study, over a period of years on a large sample, is called for, to show a more accurate picture.

References

(1)   British Standards Institute (1984) *BS 3811: 1984 Glossary of Maintenance Management Terms in Terotechnology*. HMSO, London.
(2)   Department of the Environment (1972) *Research and Development Bulletin – Building Maintenance – The Report of the Committee*. HMSO, London.

(3)   Hillebrandt, Patricia (1985) *Economic Theory and the Construction Industry*. Macmillan, London.

(4)   Hillebrandt, Patricia (1985) *Economic Theory and the Construction Industry*. Macmillan, London.

(5)   Briscoe, Geoffrey (1988) *The Economics of the Construction Industry*. Mitchell in association with the Chartered Institute of Building.

(6)   Briscoe, Geoffrey (1988) *The Economics of the Construction Industry*. Mitchell in association with the Chartered Institute of Building.

(7)   Department of the Environment, Welsh Office and Scottish Office (1993) *Housing and Construction Statistics 1983–1993*. HMSO, London.

(8)   Building Maintenance Information (1992) Slow recovery from maintenance work slump, *BMI News March*. RICS.

(9)   National Economic Development Office (1991) *Construction Forecast 1991–92–93, Summer 1991*. NEDO.

(10) Hecht, Francoise & Doyle, Neil (1991) Repair cuts will come back to haunt owners, *New Builder*. 25 April.

(11) Building Maintenance Information (1991) Storing up problems for the future, *BMI News September 1991*. RICS.

(12) Thomas, A. & Archer P. (1989) Lies, damn lies and the English House Condition Survey, *Environmental Health*. February.

(13) Building Maintenance Information (1990) *Occupancy Cost Indices – Description, Methodology and Background* Special Report 190. RICS.

(14) Building Maintenance Information (1992) Maintenance expenditure starts to decline, *BMI News March 1992*. RICS.

(15) Bibby, S. (1992) *The Economics of Maintenance*. (Undergraduate Thesis, Leicester Polytechnic).

(16) Bibby, S. (1992) *The Economics of Maintenance*. (Undergraduate Thesis, Leicester Polytechnic).

(17) Building Maintenance Information (1992) Maintenance expenditure starts to decline, *BMI News March 1992*. RICS.

(18) Bibby, S. (1992) *The Economics of Maintenance*. (Undergraduate Thesis, Leicester Polytechnic).

(19) Department of the Environment (1982) *English House Condition Survey 1981*. HMSO, London.

(20) Department of the Environment (1988) *English House Condition Survey 1986*. HMSO, London.

(21) Royal Institution of Chartered Surveyors (1990) *Building Surveyors Workload Statistics Survey 1989/90*. RICS.

(22) Audit Commission for Local Authorities in England and Wales (1986) *Improving Council House Maintenance*. HMSO, London.

(23) Audit Commission for Local Authorities in England and Wales (1986) *Managing the Crisis in Council Housing*. HMSO, London.

(24) Association of District Councils (1989) *A Time to Take Stock – Recommendations by the Association of District Councils to improve the condition of the housing stock*. ADC.

(25) Association of District Councils (1989) *A Time to Take Stock – Recommendations by the Association of District Councils to improve the condition of the housing stock*. ADC.

(26) Royal Institution of Chartered Surveyors (1991) *Britain's Environmental*

Strategy. RICS.

(27) Department of the Environment (1988) *English House Condition Survey 1986*. HMSO, London.

(28) Audit Commission for Local Authorities in England and Wales (1992) *Developing Local Authorities' Housing Strategies*. HMSO, London.

(29) Collins, J. (1986) Maintenance – key to success, *Chartered Surveyor Weekly*. 1 May.

(30) Audit Commission (1991) *National Health Service Estate Management and Property Maintenance*. HMSO, London.

(31) National Audit Office (1991) *Repair and Maintenance of School Buildings*. HMSO, London.

(32) National Audit Office (1991) *Repair and Maintenance of School Buildings*. HMSO, London.

(33) National Audit Office (1991) *Repair and Maintenance of School Buildings*. HMSO, London.

(34) National Audit Office (1991) *Repair and Maintenance of School Buildings*. HMSO, London.

(35) Audit Commission for Local Authorities in England and Wales (1988) *Local Authority Property – A Management Handbook*. HMSO, London.

(36) Building Maintenance Information (1991) School maintenance fails the test, *BMI News*. September.

(37) *Building Surveyor* (1991) Issue 2, 2nd October.

(38) Edwards, H. (1992) Local Management of Schools, *Chartered Surveyor Weekly/The Building Surveyor*. March.

(39) Franks, J. (1992) *Building Procurement Systems*. CIOB.

(40) Department of the Environment (1972) *Research and Development Bulletin – Building Maintenance – The Report of the Committee*. HMSO, London.

(41) Building Maintenance Information (1989) *Building Maintenance – Investment for the Future*. Special Report 176/177. RICS.

(42) Building Maintenance Information (1989) *Building Maintenance – Investment for the Future*. Special Report 176/177. RICS.

(43) Building Maintenance Information (1990) *Occupancy Lost Indices – Description, Methodology and Background*. Special Report 190. RICS.

Chapter 2
Maintenance Organisations

The position of a maintenance department within an organisation is dependent on the strategic objectives of that organisation and the importance it attaches to the condition of its buildings. Any study of a maintenance department must consider:

❑ Its position within the overall organisational structure
❑ The organisation of the maintenance department itself

Before either of these can be discussed in any detail it is essential to establish the organisation's attitude to maintenance, as this will influence the policy framework within which maintenance operates. This in turn will not only have a major effect on organisational matters, but also on some other facets, such as operational details, and the approach of the organisation to the procurement of buildings.

The context within which maintenance exists

Buildings may be considered to be a facility, and this requires maintenance to be viewed in the wider context of the emerging discipline of facilities management[1]. Powell[2] defines facilities management in a number of ways.

'It is concerned with the systematic optimisation of our property and use of our environment.'

'Facilities Planning determines how an organisation's tangible fixed assets best support achieving the organisation's objectives.'

The American Library of Congress gives the following, more comprehensive definition:

'The practice of co-ordinating the physical workplace with people and the work of the organisation, integrating the principles of business administration, architecture, and behavioural and engineering services.'

Powell goes on to comment that facilities management (FM) is also intrinsically linked to change, thus emphasising that buildings' management must operate in an essentially dynamic climate. Related to FM is the notion of asset management, which tends to focus management attention on those facilities or assets that are perceived as making a direct contribution to corporate profit making objectives. In many sectors of industry buildings are not viewed in this way and, thus, effectively become excluded from active consideration in terms of asset management.

The emerging role of the facilities manager tends, increasingly, to be involved with those issues perceived by senior management to be important to the efficient usage of buildings, such as space utilisation and energy consumption. Whilst this is laudable, all too frequently, there is little or no active consideration given to the maintenance of the building fabric, as this is not perceived by senior managers to be part of the efficiency equation.

A better comprehension of the impact that building fabric may have on the productive performance of the built environment is beginning to change perceptions. For example, there is an increasing understanding of the phenomenon of sick building syndrome, and this is focusing attention on the less obvious effects of poorly maintained fabric and installations.

An apparent lack of interest in fabric condition manifests itself in managerial, technical and financial neglect, which must be considered a facilities management failure. However, there are cases where the buildings are recognised as being more directly related to corporate performance. They may, for example, be a direct generator of income, in which case there is often a presumption that market economics will prevail to force owners to maintain buildings in a proper condition through rents. This assumption of perfect market conditions, however, is rarely accurate[3].

Even if it was true, there is no guarantee that this leads to a rational approach to managing building condition. Rather perversely, the response to rent pressures is likely to be a superficial one. The customer in the market place tends to be rather unsophisticated in terms of building condition, and often rather easily seduced by a cosmetic response. For this reason serious building condition problems are rarely tackled at the correct time.

Within the public sector a fresh range of problems present themselves.

The perceived importance of building fabric varies according to the use of the building, and different standards may be adopted for schools, health care buildings and local authority housing. However, many of these buildings are not seen as a generator of income, but rather as one of the means by which a whole range of society's needs are met. Although market pressures may be limited, there will be other pressures, social and political as well as economic. Almost inevitably there will be a need to reconcile conflicting demands with limited resources, which almost always militates against adequate funding being allocated for building maintenance. Parents of school children, for example, will be much more concerned that a school has adequate book stocks and that there are small class sizes than they will about the condition of a school, so long as it is apparently in reasonable repair. Local authority managers can realistically only have one option given these political realities.

What emerges from this brief discussion is that maintenance policy has to be considered not only in the light of a building's function, but also in relation to the user's perception of the building's condition, and its relevance to their primary needs. This latter effect is extremely difficult to quantify and will be discussed in more detail later. However, what is certain is that neither facilities management nor maintenance management can operate without a clear picture of what is being managed and what the functional requirements are.

Management in all areas requires the stating of objectives against which performance is judged. The corollary of this, in relation to buildings' management, is that the defining of user needs is an essential prerequisite for the evaluation of building performance. Only then does it become possible to determine what action, if any, is necessary to maintain the building's function, and to provide a benchmark against which performance of maintenance operations can be judged. Such considerations should, of course, encompass not only animate users but also inanimate ones, such as plant, equipment and machinery.

Proper maintenance management, therefore, should embrace not only managing the building in use but also play an important part in its procurement. If this process is carried out correctly then a comprehensive performance model of the building should be constructed at the outset, resulting in a skilfully developed design. At the completion of the construction phase, the building owner and the user, who may not be the same, should be in receipt of a facility together with the necessary instructions for its proper use. Unfortunately this happens all too rarely.

Maintenance policy framework

BS 3811[4] defines maintenance policy as a strategy within which maintenance decisions are made. This may be considered as a set of ground rules for the allocation of resources between the various types of maintenance action that can be taken. Maintenance policy should be considered in the widest possible context throughout all the phases in the life cycle of a building. Furthermore, it needs to be recognised that policy influences on maintenance may not always be direct ones. In other words it is possible to distinguish clearly between:

❑ Policy which is specifically directed toward building maintenance
❑ Policy decisions taken with respect to other matters, but which will influence maintenance

The attitude, or stance taken by decision makers, during the procurement of a building, will have a profound influence, not only on the strategic design, but also on the building's subsequent performance. All decisions should be carefully examined, and the possible consequences for the building throughout its economic life considered. In simple terms the building cycle can be described in the following six stages:

❑ Brief
❑ Design
❑ Procurement
❑ Construction
❑ Commissioning
❑ Operation

These are all considered in detail later, but, at this stage it is necessary to identify their relevance to policy makers, and indicate their possible implications for fabric maintenance.

(1) Brief
This phase in the building's life involves establishing a performance model for the building, as an essential prerequisite for the proper and effective management of that building, including its maintenance. The model sets a standard against which the performance of the building in use can be measured. The importance the building owner attaches to the setting up of this model is indicative of the attitude he is likely to adopt towards property and its use.

(2) Design
Building design will be subject to a policy stance at two levels throughout

the process. Firstly, a position has to be taken at the conceptual level, in terms of the type of building required to perform the function in question. This may be manifested in a number of ways; for example the budget allocated to it, the time allocated before occupation is required, a specific statement on maintenance, and the anticipated life of the building. Secondly, the development of the detailed design, which follows, should be a natural consequence of initial policy decisions put into motion at the conceptual level.

(3) Procurement

The basic requirements for the building, identified at the early stages, will require a considered view to be taken of the most appropriate procurement system to be adopted. This may have repercussions on long term performance, and hence on fabric maintenance requirements. For example, a need for early occupation may dictate a fast-track approach, which will place constraints on the design, and possibly result in consequences for fabric maintenance in the future. It is important to emphasise this causal link, and to stress that the likely outcome of these policy decisions should be analysed to their logical conclusion.

(4) Construction

The outcome of the construction stage, which is conditioned by earlier design activities, may be judged by assessing how well the building meets the client's basic requirements. Furthermore, client satisfaction will also be influenced by the quality control exercised by all parties during site operations. The analysis of building defects suggests that, whilst designers and contractors share the responsibility more or less equally for building faults, there are instances where policy makers within the client's organisation must also take some of the blame.

(5) Commissioning

The culmination of the preceding stages in the procurement process is delivery of the building. In too many cases the way in which this is performed is exceedingly unprofessional, not only in terms of administrative and practical considerations, but also in relation to the information provided to the occupier/owner on the asset he has acquired, often at great cost. The effectiveness of the handover and commissioning phase is a key determinant in the subsequent performance of the building, and improvements are only likely to come about when there is an increased awareness of this link by building owners, which will prompt them to demand a better service.

(6) Operation

The position adopted by management with respect to the occupation and running of their buildings will be consistently subjected to a range of pressures, including commercial, aesthetic, social, political and economic. The essential issue is not so much that maintenance should be given higher priority, but rather that the need for maintenance is recognised in the first place. If competing demands for scarce funds cannot all be satisfied, any decision not to fully fund maintenance work should only be taken after a carefully considered analysis.

The operational phase is inescapably linked with what has gone before. Once the building is occupied then its operation and usage falls under the influence of the organisation's policy with respect to all aspects of its mission.

At the time of occupation there are three possible scenarios:

❏ Maintenance of the new building will be carried out within an existing maintenance management framework, which represents a continuation of existing policy
❏ A maintenance management framework does not exist and must therefore be formulated, in which case there should have been early recognition of it so that it is put into place at an early enough stage to inform all stages of the process
❏ The new building is so significant, in relation to the existing building stock, that it calls for a rethink of existing procedures and systems, which should ideally have been initiated at conception

Maintenance policy issues

Whatever scenario exists, in considering the operation of maintenance management, there are a number of common areas requiring a policy statement.

Resource allocation

The proportion of resources that will be allocated to building maintenance will have to be determined in a competitive environment. These resources may be in terms of finance, staffing (both managerial and operative) and time. Generally, maintenance tends to compete on rather unfavourable terms for all of these, and for finance in particular.

Following the allocation of maintenance resources as a block, is the need to decide precisely how these resources are to be distributed. Given the inevitable pressures, this may be carried out in a variety of ways, some of which may have little to do with building performance considerations, and be beyond the influence of the maintenance technical staff. The process may be the result of a clearly defined policy or of some mysterious internal process, dictated by other characteristics of the organisation.

Performance requirements

If a logical approach to building performance has been taken from inception, then a detailed performance model may exist. This relates of course not only to technical standards, but also to operational and financial ones, such as response times and budgets. However, all too often policy in this respect has not been derived from a professional building performance standpoint.

Execution of the work

A policy will need to be formulated to indicate how maintenance work is to be executed. This will involve consideration of such factors as:

❑ Who executes the work
❑ When is it executed
❑ How is it executed
❑ How is it supervised and controlled
❑ Its relationship with other activities in the organisation

Administrative activities

Consideration of work execution requires an assessment of the procedures necessary to administer maintenance operations, and this strikes at the heart of maintenance management. The type of maintenance department may or may not be a result of a carefully formulated policy, but will certainly be a reflection of the parent organisation's attitude to the maintenance of buildings.

Position of the maintenance department within the organisation

The position of the maintenance department within the organisation, and its relationships with other departments and functions, may be the single biggest indicator of the degree of importance attached to maintenance by senior management. A carefully integrated maintenance department probably indicates a positive policy stance, where building maintenance has been considered as an important part of the organisational objectives. This is obviously related to overall corporate objectives. However, in too many instances the reverse position will be the case, which reflects the low priority given to property maintenance by many organisations.

The discussion so far establishes maintenance in the context of the organisation, and has identified in broad terms how policy may directly and indirectly affect the extent of its influence and the constraints placed on it. A large number of important issues have been touched on, the significance of which must be considered further in the context of a business framework.

The business organisation

Organisation theories

Organisation theory has considered three approaches: the military pattern, the human relations determined, and the systems controlled. There is now a wealth of reference material available, that espouses principles which, at one time or another, could be applied to maintenance organisations. (See, for example, Hales[5], Shaw[6] and Appleby[7].)

Traditionally, it was argued that organisational structures should follow the formulation of strategy and, whilst this is logical, the reality is that they evolve over extended periods of time and therefore have to be adaptive. The classical, or military, approach is characterised by a command structure that is very formal and rather rigid. Its operation depends on very clear definitions of authority, responsibility and accountability. Increasingly, there is a recognition that more informal relationships exist, and that these can be organisationally effective, if acknowledged and channelled properly. This has given rise to the human relations approach.

Latterly, the systems approach is gaining increasing attention, as it argues that the flow of information should be the major determinant of an organisational structure.

Each of these theories characterises organisational patterns, management styles and administrative systems associated with them, and although in practice it is unlikely for an organisation to fit absolutely into any one, it will tend to exhibit tendencies. At best it is only possible to comment broadly on these tendencies, as it is rare for an organisation to be designed from first principles, and the maintenance organisation, in particular, is unlikely to be the result of a great deal of forethought.

Business organisations evolve as a company grows, and will therefore be a function of the rate and nature of this growth. In this respect there is a need to be concerned not only with size in absolute terms, but also with the diversification geographically, and the nature of the output. Many companies begin from very humble origins, where management is in the hands of a small number of people. As the company expands it becomes necessary for a more formal organisation structure to develop (figure 2.1). At this point, notions such as span of control, delegation, accountability and authority have to be taken into account. This may, or may not, result in a formally designed organisation structure, with fully defined roles, responsibilities and job specifications, although modern management thinking tends to espouse a more scientific approach to the issue.

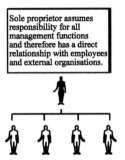

Figure 2.1 Sole proprietor.

There is now a much wider acceptance of management education as a preparation for a career in business, and a much more open ended debate about the principles, which are seen much more as being at the socio-technical-economic interface. Handy[8], for example, has equated the norms and values which determine the way in which an organisation is operated and structured to cultures. Within this he considers the following:

- Power culture, which depends on a central power source, typical of small family businesses and characterised by rapid response to change because of the low level of bureaucracy
- Role culture, where roles are precisely defined, which is characteristic of many public sector organisations, and might be considered bureaucratic in nature
- Task culture, in which people are organised in teams around specific tasks

❏ Person culture, where the individual is the central figure, the best examples are specialists who seek to operate independently within an organisation

On the other hand, some theorists embrace a system view, where the organisation is viewed as an adaptive system dependent on measurement and correction through information feedback.

These two extremes are interesting, and it is clear that many of the features of the systems view are an essential part of many modern organisations; a tendency that has been accelerated by the rapid developments in the introduction of information technology (IT).

Formal organisation structures

The organisation structure may be represented by a chart, or organogram, showing the formal allocation of responsibilities between personnel, and the relationships that should exist between them (figure 2.2). This may be backed up by a corporate plan, which sets out the general obligations and responsibilities, perhaps accompanied by a policy statement and a clearly defined set of job specifications, or descriptions for the personnel in the organisation (figure 2.3).

Organograms may be useful, in that they provide an overview of the organisation structure, and may be an aid to clarity of thought. Although many of these structures may not be the result of a planned process, but rather one of evolution, it is true that major companies often undergo reorganisation exercises that focus on basic principles. This will most frequently be part of a response to events in the activities of the business, such as expansion, merger or take-over.

It is fairly obvious that an organisational structure, as well as being historically influenced, is very much a function of the nature of the

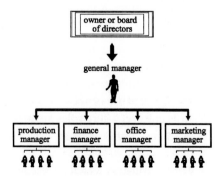

Figure 2.2 Developing organisation structure.

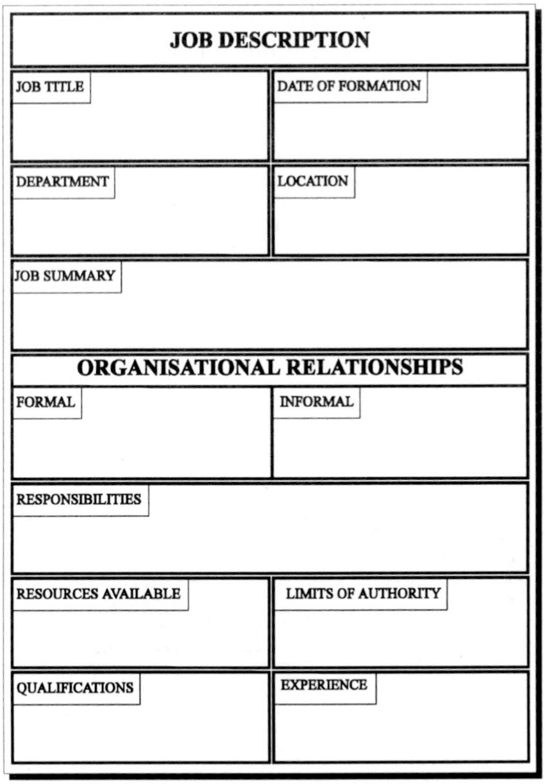

JOB DESCRIPTION

| JOB TITLE | DATE OF FORMATION |

| DEPARTMENT | LOCATION |

| JOB SUMMARY |

ORGANISATIONAL RELATIONSHIPS

| FORMAL | INFORMAL |

| RESPONSIBILITIES |

| RESOURCES AVAILABLE | LIMITS OF AUTHORITY |

| QUALIFICATIONS | EXPERIENCE |

Figure 2.3 Typical job description.

business, its objectives, and the style of management adopted. This may well be a function of policy, but the policy itself will be the product of people.

It may be, for example, that the ethos of a company favours a rigid hierarchical structure, along the lines of the classical or military pattern, and this may or may not be related to the nature of their business. On the other hand it may be that this approach is inappropriate, or that senior management endorses a different attitude, so that the organisation is designed in such a way as to emphasise the importance of human relationships.

To illustrate the confusion that may arise in passing judgements, it can be said of the organisation of much of the construction industry that it is characterised by a high level of informality, and that this is both its strength and weakness. It is also easy to identify a number of companies, operating in similar fields, and of a similar size and distribution, that have very different organisation structures and attitudes to management.

The varying approaches that can be taken to construction management, for example, are illustrated in figures 2.4(a) and 2.4(b).

In general there is little conclusive evidence in such cases to infer that one approach or the other is the most effective. There are, however, obvious distinctions that may be made between public and private sector organisations, where the context within which each of these works is quite different. Any comparison of performance, under such circumstances, will be difficult.

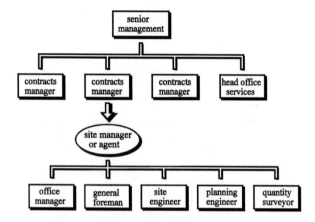

Figure 2.4(a) Contractor organisation A.

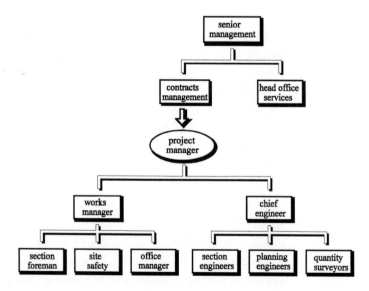

Figure 2.4(b) Contractor organisation B.

Given this range of diversity, it is not entirely surprising to find that the position of a maintenance organisation within the overall company structure varies hugely.

The building maintenance organisation

Scope of the maintenance department

In its broadest sense, the term maintenance department is used to describe the person or persons responsible for the planning, control and execution of maintenance operations. This may be wholly in-house or, as is now much more likely, it may include independent bodies, such as consultants and contractors. In considering the maintenance management systems to be used, the relationships with these bodies and the rest of the business organisation must therefore be carefully taken into account. The nature of these interfaces will influence operational methods and management systems profoundly.

In general, each of the following phases must be considered in structuring maintenance departments:

❑ Generation of maintenance work
❑ Execution of the work
❑ Control of operations
❑ Provision of feedback
❑ Financial control
❑ Evaluation of performance

The department set up to deal with maintenance needs must address two major concerns. Firstly, it must provide an appropriate service within the guidelines established by proper consideration of corporate objectives and, secondly, it must be capable of judging its own effectiveness by monitoring and controlling its performance.

The need to satisfy these two interlinked issues underlines the importance of the interface of maintenance with the rest of the organisation. For example, a typical housing association defines a set of performance indicators, amongst which are the following:

❑ Accountability
❑ Response to repairs
❑ An indication of the times properties are unoccupied

Allied to these, it sets clear performance requirements for maintenance operations, as an integral part of satisfying a set of target performance indicators.

The organisation of the maintenance department will normally be determined by the characteristics of the parent organisation, except in cases where fabric maintenance is essential to corporate objectives, for example in a housing association. In this case the maintenance department could have a strong influence in determining the nature of the overall organisation.

Types of maintenance department

It is possible to produce a very generalised classification of maintenance organisations depending on the degree of domination exerted by one or other of the parties involved in the process. Using this approach four main parties are identified:

❑ Occupants or tenants
❑ The owners or client organisation responsible for managing the property
❑ The 'professional' maintenance team
❑ The maintenance work force and their immediate supervision

Each party is perceived as likely to follow different, and perhaps conflicting, goals. Assuming that organisations may sometimes be dominated by one party, four logical types of departmental organisation are thus derived.

(1) Occupant dominant type

The major emphasis here is on speedy service to occupants where work is initiated by occupant requests. As quick service is required, the emergency response system becomes overloaded, resulting in highly unpredictable work loads, and probably an inefficient use of resources. Management is concerned with keeping good relations with occupants, and control systems are related to speed of service rather than productivity. Costs are difficult to predict, but tend to be relatively high through the need to maintain large labour strengths capable of meeting peak demands. The labour force will tend to be in-house.

(2) Owner/client dominant type

The major goals here will be:

❑ Maintaining the value of the property
❑ Keeping costs as low as possible
❑ Ensuring that properties are let or utilised as soon as possible

Work tends, under this regime, to concentrate on external painting and repairs during occupancy, and internal refitting and decorations at change-overs. Management therefore develops a system of maintenance planning related to occupancy patterns, rather than building fabric needs. Frequently, contract labour will be used, due to extremely variable work loads. Thus management will focus heavily on contract control and competitive pricing.

(3) Professional dominant type

When this group dominates it is most likely that the style of management will reflect a sympathetic attitude to the maintenance needs of the built fabric. For example, there is likely to be a strong emphasis on planned preventive maintenance programmes, which are seen as limiting the amount of random or emergency work through a proper care regime. If applied rigidly, control will be based on achieving quality, and in some cases may result in unnecessary work being carried out. The work force will probably be a highly trained one, working within a carefully laid down pattern.

(4) Work force dominant type

Maintenance work, in this case, is influenced largely by operative work groups and their immediate supervisors. Standards will be heavily affected by trade group norms, and there will thus be large variations in the quality of work. Control and management will be a grass roots affair, with a very minimal professional organisation, and thus low overheads.

Within each of these, characteristics of Handy's cultural approach are clearly apparent. Whilst not a very effective classification basis, the foregoing categories are useful in terms of what they disclose about maintenance organisations. Another approach that can be taken is to classify a maintenance department in terms of the characteristics of the building stock and the nature of the organisation, including the importance attributed to maintenance within it. This gives rise to three variables.

(1) The importance attached to fabric maintenance

This can be considered in four categories:

- ❑ Category 1 – primary importance, e.g. housing
- ❑ Category 2 – secondary importance, e.g. commercial with owner/occupier
- ❑ Category 3 – tertiary importance, e.g. commercial rented
- ❑ Category 4 – peripheral importance, e.g. industrial

(2) The structural characteristics of the organisation

The structural characteristics of organisations are extremely varied. There is clearly the world of difference between servicing the maintenance needs of a large concentrated estate and dealing with a geographically, or divisionally dispersed one.

(3) Characteristics of the building stock

The characteristics of the building stock can be described by each of the following:

- ❏ Age
- ❏ Type
- ❏ Condition

In structuring a maintenance department, the degree of homogeneity displayed by the building stock will be a significant factor. Where, for example, an organisation's building stock is heterogeneous, the decision may have to be made to sectionalise the maintenance effort, whereas with a homogeneous stock the need for specialist divisions is unlikely to be necessary.

The two variables relating to characteristics of the organisation and the building stock can be combined to give the following categories:

- ❏ Concentrated homogeneous
- ❏ Concentrated heterogeneous
- ❏ Geographically dispersed homogeneous
- ❏ Geographically dispersed heterogeneous
- ❏ Divisionally diverse

Functions of a maintenance department

Advisory function

This can be seen as a key area of interface, involving liaison with owners, and consultation with senior management, to advise on such matters as:

- ❏ The development of the design brief for new buildings, their design and procurement
- ❏ The production of as-built drawings and maintenance manuals
- ❏ The performance requirements of new buildings in general
- ❏ The provision of specialist advice, and other services related to the areas of adaptation, refurbishment and extensions/modifications
- ❏ Determination of standards to be achieved, and the setting of

performance indicators in relation to the primary needs of the organisation, e.g. quality and response times

❏ Providing on-going information on building condition, which in conjunction with financial information, may help senior management in budgeting decisions, and also on decisions whether to repair, replace or renew

❏ On-going information relating to maintenance costs, to assist in sensible financial management

❏ Advising senior management on the organisational needs of maintenance, to ensure that an efficient organisation exists, with the correct relationship to the rest of the organisation

Organisational function

This must be considered with respect to internal functions and, also, with points of interface, so that each of the following may be relevant.

(1) The formation of a basic internal administrative system that clearly defines:
   ○ Roles and responsibilities
   ○ Organisational inter-relationships
   ○ Communication channels
   ○ Chains of command and patterns of accountability
   ○ Standard procedures

(2) The defining of proper protocols for dealing with external organisations such as contractors and consultants, and other departments within the organisation. Within this function careful consideration will need to be given to the procedures for communicating information, whether written or verbal. Increasingly, information technology is of critical importance when considering administrative and organisational systems.

Operational function

The relevant operations can be classified under the following headings:

❏ Work input
❏ Programming the work
❏ Ensuring the work is executed
❏ Monitoring and controlling quality, cost and time
❏ Authorising and arranging payment
❏ Provision of management information including feedback

In organisational terms, these operations represent the essential maintenance execution function which, with the detailed internal organisational aspects of maintenance departments, will be dealt with later.

The typical department structures discussed below, are all summarised by the use of an organogram. This formal chart does not always exist, and in many organisations it is informal working relationships that ensure higher levels of service. In other words, there are legitimate situations where the sacrifice of formal control may better serve overall corporate objectives.

Typical maintenance organisations

Local authority property maintenance

The Audit Commission for England and Wales have published a number of reports which present a concise picture of the nature of property management in the local authority sector. The most relevant reports in this respect are two related publications, concerned specifically with management. (*Local Authority Property – A Management Handbook*[9] and *Local Authority Property – A Management Overview*[10]).

These reports advise on wider property management considerations, but provide some information from which some useful conclusions can be drawn with respect to maintenance management. Both reports exclude housing maintenance, presumably because housing represents a particular set of problems, and needs to be considered in relation to the overall provision of what is termed social housing, which is not restricted to local authority properties.

The Audit Commission divided the local authority property portfolio into three categories.

(1) Property held for the direct provision of services
Property occupied by local authority staff for the direct provision of its services, including elderly person's homes, schools and offices, leisure centres, police and fire stations.

(2) Tenanted property
Property, which provides services in an indirect manner, for example, industrial starter units or community facilities, and which may also receive local authority funding in the form of a subsidy. Also included under this category would be commercial property held as an investment, which is expected to provide a commercial rate of return, for example shops and offices.

(3) Vacant property

Property in the ownership of the local authority, surplus to requirements; property occupied for a temporary purpose, in advance of the main purpose for which it was acquired; and property that has become redundant.

The mix of properties falling into each of these categories within a given local authority varies with policy and there are also major differences due to geographical factors.

The issue of school maintenance is a particularly serious one, but is not given separate treatment here, because in organisational terms, and in line with Audit Commission guidelines, it is managed within the wider property portfolio perspective. At the time of writing, the issue of school maintenance is also complicated due to the move towards local school management, with every headteacher becoming a budget holder.

The reports identified a series of management problems that all stemmed from what they considered to be an inadequate strategy for managing property. These problems are listed below, and to some extent they may all impinge on maintenance performance. The two that are highlighted however are of prime importance.

- ❏ **Inadequate management information**
- ❏ No incentive to users
- ❏ Failure to carry out property reviews
- ❏ Opportunity cost of holding property not recognised
- ❏ Confused objectives for tenanted/vacant property
- ❏ **No co-ordinated maintenance policy**

It has to be said at this point that there is a general assumption of underfunding for maintenance work in all areas of the public sector, but this does not preclude consideration of the efficiency with which such resources are utilised.

In order to solve these problems a series of recommendations were made which have implications for the organisational strategy that should be adopted. In the first place there was a strong recommendation for all authorities to have a written policy statement with respect to property management, and that maintenance of the building stock should be part of this. To implement this policy, there was also a suggestion that authorities should set up a property committee, sub-committee or equivalent body, to determine an effective strategy for managing the resource or asset (figures 2.5 and 2.6).

Property management responsibilities should also be clearly defined for members, staff and building occupiers. The reports stated that every

Figure 2.5 Defining responsibilities – an overview.

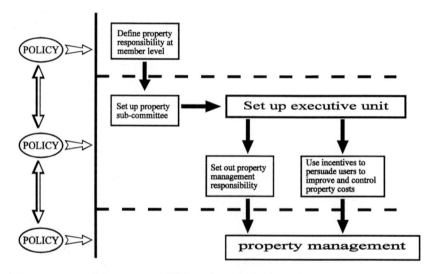

Figure 2.6 Defining responsibilities – local authority.

authority should have a full property record, and this should include the condition of the buildings. From this a five year maintenance plan should be prepared, taking account of the age profile of the stock. They insisted that a properly constructed costing system should be set up and, for this reason, maintenance costing in local authorities tends now to be part of an integrated management system.

Figure 2.7 shows the maintenance management structure within a shire county council, from which it can be seen that the council has a property department with a director and a number of assistant directors, each responsible for an area of property management. The major interest, in this example, is with the assistant director for property maintenance, whose brief extends somewhat beyond maintenance.

Historically, maintenance managers had provided a service to a series

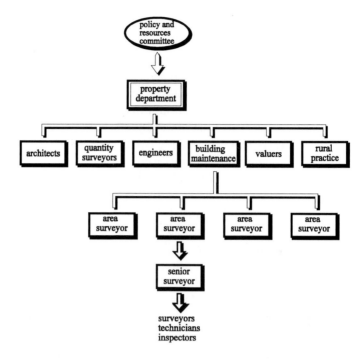

Figure 2.7 Local authority property management.

of client committees[11], which led to poor control of expenditure. For example, the needs of social services for maintenance work differed very greatly from education, and inappropriate allocation of resources tended to feed the development of backlogs of maintenance for some building types, whereas other maintenance budgets might be under-utilised.

The setting up of an integrated property department was accompanied by a change in the way in which finance was distributed. The assistant director for maintenance receives a maintenance budget from the Policy and Resources Committee of the council, and he then has responsibility for deciding how it should be spent. Since being set up the department, illustrated in figure 2.7, has carried out condition surveys on all their properties, and this provides the basis for directing expenditure in the most effective way. Individual services within the council may find this approach contentious, but it is consistent with Audit Commission recommendations, and also has the advantage of simplifying the political pressures to which the maintenance department may be subject.

The maintenance organisation is divided into four areas. The initial tendency, when the department was being set up, was to make a simple geographical division, based on the number of properties. The assistant director was, however, anxious that each area should have approximately the same demands placed upon it, so that resource allocation and the defining of staff responsibilities could be simplified.

The property records had been consolidated into a computer database, which included some limited condition data. Each property in the database was allocated a points score, which depended on the building type, its age, size and condition. The geographical areas were then delineated in such a way that the total points score of all the properties in each area was roughly equal. This was carried out on the basis that a total of 10 000 points were available and the aim was for each area to have 2500 points, plus or minus 5%.

A major relationship exists with the Treasurer's Department for the purpose of financial control. The accounting system is a highly centralised one, which is typical of local government, and maintenance cost control works within it.

The assistant director is satisfied that the organisation works well, within the resourcing constraints imposed on it.

School maintenance

The major part of the above department's building stock consists of school buildings, and the organisation has had to adapt to cater for the move towards local management of schools. This has been addressed, as in most LEAs, by using a landlord/tenant arrangement. The local budget holder is considered to be a tenant, and the property department the landlord. Maintenance is then divided into that for which the tenant is responsible and that for which the landlord is responsible. In other words, the school may themselves arrange for the execution of certain items, but for planned and preventive maintenance of the structure and fabric of the school buildings responsibility is retained at the centre.

In general the council retains about 65% of the maintenance budget in order to carry out its landlord duties. This also raised another issue. Traditionally, schools had been funded purely on a simple pupil numbers basis. This approach was not considered to be equitable for the allocation of a maintenance budget, as no account was taken of other factors, such as the age of the school and its condition. Past maintenance and condition survey data allowed the schools to be classed in terms of condition, which then assisted the maintenance department to rank the schools in terms of the amount of money it considered should be spent on each, subdivided according to whether maintenance was a landlord or tenant responsibility. This provided the basis for a five year planned maintenance programme for the landlord portion of the maintenance responsibility.

The tenant portion of these forecasts was then further analysed for the following five years to produce a weighting factor for each school, which was applied to the actual number of pupils in each school to give a weighted school population. The budget allocation to each school includes a portion for maintenance, based on this weighted population.

Another approach that has been adopted elsewhere[12] is to take the actual gross floor area of a school and multiply this by a condition indicator varying between 1.0 and 5.5, according to the results of condition surveys. This gives what is termed an adjusted floor area, and allocation to schools is based on it.

There is, of course, no guarantee that the locally managed portion of the maintenance budget will in fact be spent on maintenance, an issue which causes some concern amongst property professionals.

Housing maintenance organisations

As the National House Condition Surveys indicated, housing maintenance is a major problem throughout all sectors, including the owner occupied one. Generally, however, the majority of data available represents what we may term social housing, with the two major providers being local authorities and housing associations.

Housing maintenance has also, due to its high political profile, received a major share of the attention devoted to maintenance management, and in particular to the application of computer based systems. As well as its political significance, this sector also has structural characteristics, in particular the homogeneous nature of its stock in terms of building type, and a relatively limited geographical distribution of its stock within a given authority. Divergence from this simple model raises serious problems. For example, some authorities and agencies have aged housing stock, and many housing associations have a large proportion of refurbished property. Both of these situations cause specific maintenance problems, and may influence the organisation of maintenance operations.

There are a number of other general problems faced by housing management teams in the public sector, including those related to the provision of sufficient funding. This may be a question of total funding, or the way in which allocations are made, which identifies the potential for a major conflict in terms of objectives.

Local authority housing

The general difficulty with public sector housing maintenance is that, although there may be general agreement on the objective of keeping the housing stock in a good state of repair, there is no one clear definition of what this condition might be, notwithstanding the provisions of the Housing Acts. There will be different priorities emphasised by tenants, housing managers, maintenance personnel, surveyors, supervisors and operatives.

Tenants, for example, are likely to be concerned with those repairs that seem of most direct relevance to them personally, and naturally want those repairs carried out as soon as possible. The maintenance professional, on the other hand, may prefer a system of planned maintenance that reflects a long term attitude to maintaining the fabric of the building. The housing manager will be interested in housing as a social service and, when resources are apportioned, the conflicting demands of different types of tenant may be a stronger influence than condition factors.

Public sector housing has seen many changes in recent years, and constant political pressures are the norm. There have been substantial moves away from local authority control and greatly increased activity in low-cost rented housing from the housing association movement. This trend may or may not continue, as it is subject to political pressures, which may change from time-to-time.

The increasing diversity of low-cost rented housing provision highlights interesting organisational factors. It is possible to identify organisations which are owner, occupier, professional or work force dominant systems.

One of the major stated objectives of the diversification of provision has been that of breaking down large bureaucratic organisations into smaller units, on the basis that this provides shorter lines of communication, i.e. the possibility of a more personalised service. The attractiveness of this was recognised some years ago, and resulted in attempts by large local authorities to decentralise operations, and use area based work groups as rapid response teams. This can, arguably, be perceived as a marriage of an occupant and work force dominant organisation. The approach has its roots in attempts to counteract perceived social problems and tenant satisfaction, to which housing conditions represent one of many contributory factors. This is an example, quite commonly encountered, of housing management taking a stance which may have more to do with social factors than real fabric maintenance.

The accepted wisdom amongst maintenance professionals now, however, is for the introduction of planned maintenance programmes, and organisational structures reflect this. The benefits of planned programmes were stressed by the Audit Commission in 1986[13], which strongly criticised housing maintenance management, and, whilst accepting a possible underfunding, considered that there was excessive waste. The report argued very forcibly that the adoption of more carefully organised planned programmes could bring about major improvements in the efficiency with which resources were utilised.

The 1986 report went on to give a series of recommendations amongst which were the following.

(1) An integrated organisation should exist, with overall responsibility for determining maintenance plans, priorities, customer service standards and for monitoring performance, clearly fixed within the housing department. This recommendation is, however, somewhat in conflict with other arguments, supporting a decentralised approach on the grounds that, although costs may rise, service quality improves[14].

(2) The execution of jobbing repairs should be optimised by careful consideration of the problems faced, and by adopting an organisational strategy to tackle them. For example through the use of:

- ❏ Estate based repairs
- ❏ Zoned maintenance
- ❏ Neighbourhood term contracts

(3) Clear service requirements should be established, and performance monitored through an effective management structure.

These recommendations, together with other factors, account for local government housing maintenance being treated separately from the maintenance of other public buildings.

Housing associations

Housing associations represent a diverse range of non-profit making private organisations, with a clear mandate to provide low cost rented housing. The responsibility for their operations is vested in a management committee, and their origins lie in so-called voluntary housing movements. They vary enormously in their size and type of provision but, in general, operate on a smaller scale than local authority housing departments. Most of them are based in confined geographical regions, where they operate independently of both local and central government, although they are frequently in receipt of substantial public funding through the Housing Corporation, which supervises their operations, and may also act in an advisory role.

The *Committee Members' Handbook*[15] states that:

'the Committee needs a clear policy on both current and future maintenance and repair. Such a policy must take into account the needs of future tenants as well as those of the present, and under the new Housing Association Grant regime this will include making provision for sinking funds to cover the cost of infrequent and possibly unexpected repairs.'

The National Federation of Housing Associations (NFHA) produce a number of guides for members and employees. Their publication, *Standards for Housing Management*[16], for example, provides an easy-to-use guide to good management practice and the formulation of policy. On maintenance organisations it says that their policy should be set out so as to:

❑   Ensure that statutory and contractual obligations are met
❑   Provide a responsive and effective service to the client
❑   Maintain capital assets
❑   Ensure that repair and maintenance is cost effective

These are rather sweeping guidelines which are supplemented, elsewhere in housing association literature, by further recommendations regarding the need for their operation to:

❑   Strike a balance between landlord and tenant's responsibilities
❑   Provide a service which is customer centred
❑   Give good value for money
❑   Ensure a consistent operation of policies and procedures
❑   Be subject to sound budgetary control

The annual reports of most housing associations will include comment on the condition of their stock, and some assessment of their maintenance performance.

Because of their diversity, and despite clear guidance, there is great variation in performance levels, which does not always appear to be related to their organisational style. They do tend, however, to be characterised by a high level of commitment amongst their staff and, because of their smaller scale, have shorter lines of communication, which usually results in a closer informal linkage between management and tenant.

The ethos of the housing association movement has also directed much of its attention to social priorities, such as inner city programmes, where it enjoyed access to grants for the major repair aspects of rehabilitation programmes. At the time of writing the funding arrangements for major repair programmes are being subjected to substantial changes, following the 1988 Housing Act. A report by the NFHA[17] quotes:

'The post 1988 regime passed to associations "maintenance risk" in the form of responsibility for the cost of longer term maintenance and major repair.'

Under the Act, associations will no longer have access to major repair grants, and are expected to make provision out of rent income. Initially, with the help of the Housing Corporation, associations set this provision at 0.8% of stock replacement value for new build schemes and 1.0% for rehabilitation. In the case of the former they estimated that this was equivalent to £11 per week additional rent.

The NFHA report was concerned on a number of fronts. In the first instance the thrust of the 1988 Act was towards affordable social housing. Their research indicated that some provisions of the Act militated against this, and was leading to a shift of intervention from grant to rent subsidy.

Also of great concern was the effect this was having on the content of housing association activity. The report concludes:

'There has been a dramatic decline in both the number and proportion of new schemes accounted for by rehabilitation from 50% before 1989 to less than 20% by 1991. The switch to new build has been accompanied by a shift in development work out of the inner cities.'

There is also a fear within the movement that increasing commercial pressures will lead to more widespread mergers and take-overs and that this will negate the advantages in terms of tenant service that associations have enjoyed.

Figures 2.8(a) and 2.8(b) illustrate the organisation of two housing associations. They may both be described as medium size organisations

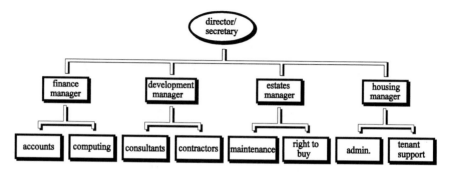

Figure 2.8(a) Housing Association A.

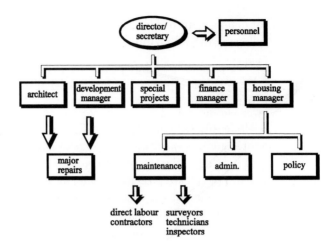

Figure 2.8(b) Housing Association B.

within their sector, having between 750 and 1000 housing units. Association A has a housing stock which is 90% purpose built, and which may loosely be termed as new-build. Association B, however, has a predominantly refurbished stock. They both operate in similar sized population centres. Whilst A has a more geographically diverse stock, with a significant number of properties in semi-rural locations, the other association works almost exclusively in an inner city environment. These differences, particularly in the nature of their stock, mean that many of the problems they face are rather different.

Association A, as can be seen, has an arm of its organisation specifically dedicated to maintenance. The section is managed by an estates manager, and is responsible for the maintenance function for buildings and infrastructure. There is a well developed cyclical maintenance programme, with a comprehensive computer based property file. This database provides an important tool for day-to-day maintenance operations, including emergency repairs. It does not at present relate to major renewal programmes.

Association B has a maintenance department that is under the control of the housing manager, but with important links to the architect and development sections. The operation of maintenance is characterised by a strongly informal set of relationships. Property records are not computerised, but to quote an employee, 'the maintenance manager is a walking encyclopaedia'. There is, however, a strong feeling that major changes will be necessary, if only due to the increasing requirements for housing associations to become more self sufficient in financial terms. The *ad hoc* treatment of property records, for example, will not be adequate for future needs. Their major problem, however, is the need for a proper condition survey of their predominantly refurbished stock. Despite this, they do not employ a maintenance surveyor. They do, however, have their own architects and employ some direct labour.

Association A employs only contract labour, and uses consultant architects for all their design services, but they do have a significant maintenance management group, which is singularly lacking in B.

There is here a clear demonstration that the needs of an organisation and its structure are not always matched. It would be tempting to comment that, if the organisation charts were reversed, then there would be a much closer matching of needs and provision.

In both cases the organisations, as they exist, are the result of historical factors and in particular the influence of personalities.

Comparison of their organisational structures raises an issue with respect to maintenance, which is clearly a central concern of housing associations, and is the subject of much debate in housing management circles. This relates to the question of where the maintenance function should be located in the organisation[18]. Association B considers that, because of its central importance in terms of tenant affairs, it should be an integrated part of housing management. Association A, on the other hand, consider that this approach does not work, as it marginalises maintenance activity. It is, therefore, convinced that a separate maintenance department, with strong management, reporting straight to the director, gives maintenance central importance and status that is clearly on the same level as other departments. There is an implicit understanding that all departments will compete for funds but should be able to do so on an equal footing.

The NFHA explicitly recognise the possibility of these two approaches to maintenance organisations, and suggest that integration of the maintenance function under the housing manager is probably effective in small associations but, that in larger ones, greater consistency is achieved by a stand alone maintenance department. They also argue that this provides a better environment for co-ordination and technical expertise, and ensures that planned work is more effective.

Maintenance in the National Health Service

Health care buildings generally represent one of the most complex building types in terms of maintenance, due to their high performance requirements, and the complexity of the engineering services needed to sustain proper levels of patient care. Like the management of school buildings, at the time of writing, there is some level of uncertainty due to the move towards the establishment of hospitals as self-governing trusts. This was causing concern and uncertainty with respect to funding levels, fuelled by increasingly acrimonious political debate. Some of this debate has related to the use and disposal of NHS assets, which includes buildings, and hence may have some influence on maintenance activity.

The change in thinking, characterised by central government guidance notes, is a reflection of the findings of a number of inquiries into health service management. For example, the Griffiths inquiry[19] stressed the importance of the property management role, as opposed to the existing functional management approach which, by implication, stresses that the estate must be exploited, and considered as a potential contributor to NHS funds.

The NHS has, over the years, been the producer of a large range of sophisticated maintenance management systems, many of them computerised. However, the software used has a decidedly estate management flavour, and is increasingly influenced by commercial thinking.

Figure 2.9 shows the general organisation of an area health authority, and figure 2.10 the management structure for a large hospital within it. At the time of writing this hospital was in the process of a move to trust status. The organogram indicates a fairly typical situation, where the maintenance manager is responsible for engineering and buildings, with energy increasingly being treated as a special issue.

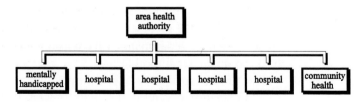

Figure 2.9 Organisation of an area health authority.

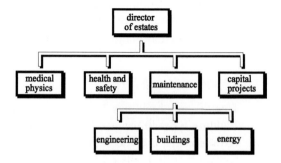

Figure 2.10 Hospital organisation.

The work of the maintenance department, at this time, divided into four main categories:

❑ Essential cover items receiving maximum priority
❑ Breakdowns
❑ Planned preventive maintenance
❑ Small scale new works and improvements

The first three categories were funded from one budget, allocated on an annual basis, with a small separate budget for the fourth item. Planning was, in general, considered to be extremely difficult, largely, it was felt because budgets were based on historical expenditure patterns, with virtually no consideration of building need.

Maintenance management within the health service is largely centred around a sophisticated computer based system, which generates planned maintenance programmes. The planned maintenance programme within the hospital studied was driven almost completely by the need to meet statutory provisions, and the requirements of insurances, and is dominated by work on plant, equipment and engineering services.

The production of accurate estate records, including building condition data, had only recently been afforded any sort of priority. Government guidelines required each potential self-governing trust to produce what is best termed a capital inventory. This includes buildings, their contents and the general condition of all assets. The initial emphasis was being given to engineering equipment and its condition, due to its high capital value. The building condition data collected was for specific estate management purposes, rather than maintenance management, although it represents a considerable improvement on the information previously held.

To produce and record this property data, the health authority under consideration was making extensive use of CAD systems. In some cases full measured surveys were being carried out, but in others, the latest available drawings were being digitised or scanned into a CAD file. It was too early to make any judgement as to whether the data being collected would provide any useful long-term benefit to the maintenance management team.

The maintenance manager was very adamant that he was underfunded. He had estimated a backlog of maintenance requiring the expenditure of £4 million over the next five years. He had actually been given £114 000 for each of the next two years, and £400 000 and £500 000 respectively for the two years following this. He anticipated that the value of his maintenance backlog would have increased substantially at the end of his forecast period. It was too early to anticipate the consequences for building maintenance of the expected move of the hospital to self-governing trust status, where funding arrangements were likely to change drastically.

Figure 2.11 illustrates the organisation of maintenance within a large Midlands hospital operating under trust status, where subtle differences should be noted. In particular, separate engineering and energy management departments report direct to estates management, rather than through the head of maintenance.

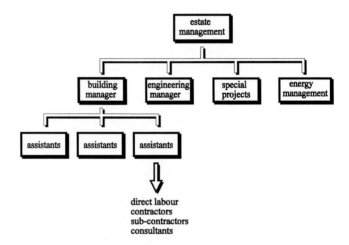

Figure 2.11 Trust hospital organisation.

Fabric maintenance in industrial organisations

(1) Company A

The company, whose organisation structure is summarised in figure 2.12, is divided into several divisions in the UK. It is part of a large diverse company, operating on a multi-national basis. At the time of writing it had just undergone major restructuring, and was experiencing major pressures due to economic recession. It can be seen from the diagram that the fabric maintenance function is under the control of a manager of estate services, who reports to the production manager of the machinery manufacturing division. This division represents the mainstream activity of the company. The estate services section is also responsible for maintenance of buildings and estates for the other divisions.

Geographically, within the UK, the company is centralised in the Midlands, where all the divisions share the same site. There is a strong policy that requires each division to be financially independent. Despite

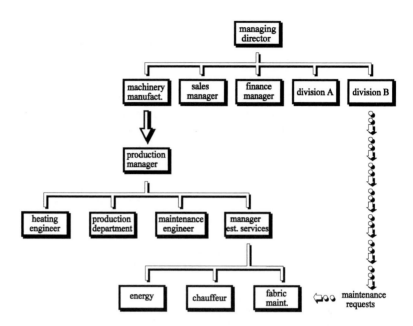

Figure 2.12 Company A.

this, there are centralised sales and finance groupings, serving all the manufacturing divisions. Similarly, for operational purposes, the fabric maintenance service can be considered centralised.

The organisation chart suggests that the maintenance function is afforded this low status by placing it under the control of one of the divisional managers. In practice, there are also complex lateral and highly informal chains of communication. It is debatable whether this can be effective, and the maintenance section felt very strongly that there was a lack of integration of their function.

There is no planned or preventive maintenance whatsoever, and the section responds to individual requests. This effectively eats up most of the budget for building maintenance. The head of the section clearly considered that they were underfunded, and there is ample evidence to support his view. In terms of corporate objectives, fabric condition has low priority in comparison to production.

The section is responsible for building fabric, site infrastructure and building services, maintenance requests being logged into a computer and given a priority rating on a numerical scale from 0 to 9. There are no systematised rules for allocating priority, which appears to be largely determined by who makes the request, although there is a clear under-standing that items which directly relate to production should have high priority. Most of these items are building services related, power in particular. This is not surprising in terms of the company's corporate objectives.

Because of the importance attached to keeping production running, each of the manufacturing divisions have their own dedicated maintenance engineer, who is not responsible to the manager of estates services. Another high priority item relates to building condition at the customer interface. Repairs required to finishes, for example, in areas where prospective clients are received, are treated with some urgency. One consequence of this is that much of the section's limited budget is spent on items that the manager considers to be cosmetic.

The manager has been with the company on its main site for 47 years, and is intimately familiar with the buildings and services for which he is responsible. There were many problems when he was absent, as there are no substantive records in terms of drawings or schedules for his building stock, which consisted of more than 40 buildings, some of them dating back to the 1920s. The newest buildings are approximately 12 years old. Most of the buildings are standard industrial structures, but there is a substantial framed office building and a large 1930s multi-storey industrial building. An established concrete cancer problem is evident, but this is not receiving sufficient attention and will only get worse.

As is typical in this type of organisation, the individual responsible for maintenance was extremely committed and constantly bombarded senior management with requests for funds he considered necessary to achieve, what was for him, a minimum standard.

At the time of writing an economic recession was imposing severe trading pressures, and over a period of three years the size of his section had been reduced by two-thirds. Fabric maintenance accounted for less than 1% of turnover. This is a common characteristic of industrial organisations.

Maintenance is, in general, carried out by in-house staff, although outside specialists often have to be employed for difficult items. In the case of some large items, company policy is that tenders should be obtained.

Another of the manager's functions was energy management, and he had installed a computerised energy management system, which, over a period of five years, had helped to bring about a 50% reduction in energy usage in real terms. Because he was able to clearly demonstrate the returns from this investment in terms of hard cash, he received strong support for this area of his work.

From figure 2.12, it can be seen that management rests mainly in the hands of the manager and his general foreman. Prior to the previous cost cutting exercise there were two assistant managers and four foremen.

The head of section considered that this was inadequate. The senior management and customer chauffeur, and a garage mechanic, also fall under his control. It can be concluded that this example is by no means unusual, and is a reflection of corporate priorities. Only time will tell if this policy is short-sighted. What is certain, however, is that a time will come, quite shortly, when the senior management will be faced with some very awkward decisions, and may well be forced into a major investment of funds. The amount required is obviously growing at an increasing rate, and what may be more problematical is the extensive disruption that will ultimately be caused to their main corporate objectives. There had already been, quite recently, a complete shut-down of a major manufacturing facility, due to an electrical failure that could clearly have been avoided by routine maintenance, if the section had had the resources to carry it out.

(2) Company B

Company B, which has an annual turnover of £90 million, is a constituent company of a major multi-national, high technology, engineering concern, based in the East Midlands. It occupies a site with an area of approximately 25 hectares, containing more than 80 buildings, with a total floor area of $100\,000\,m^2$, dating from the 1940s. The organisational structure is shown in figure 2.13.

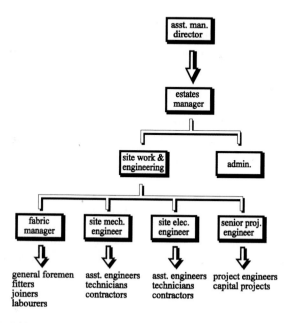

Figure 2.13 Company B.

Unlike the previous case, the estates manager here feels that his function is given relative importance. As can be seen from figure 2.13, he reports directly to the assistant managing director, has complete responsibility for maintenance including engineering, and has a staff of 41. It will also be seen that he is responsible for capital works.

His work falls under four main headings and he budgets accordingly.

(1)   *Minor maintenance*

This is work which he defines as having a low level of material usage, and which is carried out on demand by his own labour. A computerised record keeping system has existed for some time and is based on accounting procedures, from which the estates manager is able to formulate a budget for this area of work annually. These records also give him enough information to predict the distribution of expenditure throughout the months of the year, and thus allow him a measure of financial control. The current expenditure under this heading runs at around £50 000 per annum.

(2)   *Planned maintenance*

Under this heading he organises the execution of a planned programme of routine work and preventive inspection and maintenance. This is driven, to some extent, by the need to comply with statutory and insurance provisions. It gives a programme of work executed on a range of time cycles, much of it related to engineering services and plant. Other cyclical work is identified from maintenance histories and previous routine inspections. The work is executed both by direct labour and contract maintenance, the latter being on the basis of schedule of rates contracts. Budgeting is reasonably predictable, given a well established record keeping system.

Planning is also well established, through past records, and a known need to execute certain tasks at pre-determined times in the year. As an example, electrical sub-stations are always given regular inspection at Easter, to minimise disruption to production.

Under this heading are also a series of larger one-off programmes, identified from time-to-time, that have less to do with deterioration than the changing nature of the company's activities. An example of this is a complete inspection and overhaul of the site electrical systems, the original design of which was based on operations that are no longer carried out. The company operates in a rapidly changing, and highly competitive international climate, and an important part of his role is to try and respond to it.

The current annual budget for routine maintenance is approximately £200 000, although the overhaul of the electrical system involves an expenditure of £500 000, spread over five years. There is no indication how this affects resourcing of other activities under this heading.

(3) *Major maintenance*
Under this heading fall any other items of a large scale nature. Whilst the estates manager's funding requests under the previous headings are, in general, met in full, this area of work often represents things he would like to do, but which are not all supported. He receives a fixed budget for all sections of his work but in this third category all the items are given a priority rating from 'essential' to 'if affordable'. When unforeseen emergency items occur, low priority items are dropped from this budget. It thus provides his essential flexibility for the management of his total maintenance budget.

The content of this section of his work was continually under review, and he sought, where possible, to move what he considered to be very important items to routine maintenance. For example, painting had in previous years been part of the major maintenance section, but for the next and subsequent financial years, was being switched to routine maintenance. He justified this to his managing director on the grounds of more effective planning and economies of scale efficiency improvements.

The current budget under this heading is £300 000, which is the largest for any section, suggesting a substantial level of contingency thinking. The total budget for which the department is responsible under these three headings is therefore of the order of £1 million, about 75% of which can be classed as maintenance.

It is worth commenting, however, that the head of this department considered maintenance in a far wider context, and saw all his work as being related to maintaining the site in the optimum condition to help meet corporate objectives, and that refurbishment, conversion work and other capital projects were an integral part of this. Whilst on face value maintenance budgets were fixed, the largest element in their make-up falls under the major works section, which is considered to be completely flexible. This provides the potential for response to changing conditions, which is necessary in all maintenance work for this type of industrial organisation.

(4) *Major/capital works*
This is the fourth area of work which falls under the control of the

estates manager, and the annual expenditure is, as is to be expected, variable, normally in the range from £350 000 to £500 000. It is interesting to note, however, that the estates manager is seen as a major generator of proposed capital projects from his overall role as estates co-ordinator, and the comparative position of strength from which he appears to work.

The organisation was characterised by extremely supportive senior management, and an excellent personal working relationship between the estates manager and the assistant managing director. There was also an extremely effective relationship between the estates department and the accounts department. The latter operated a computerised cost accounting system, within which maintenance was fully integrated. The system was designed in such a way that constructive maintenance data could be extracted. The essential components of the system were a logical system of numbering all orders, whether for direct labour or contract execution, and the recording of this number on all time sheets and invoices. The secret to the successful operation of the system appears to be the ordering system.

The data generated provided a historical basis for budget fixing, which means that there exists a tendency for maintenance programmes to be based on past expenditure rather than on need. There were property records stored in a database format, and record drawings which were CAD based. There appeared to be no systematic attempts to carry out full condition surveys.

Despite this, the head of department felt that he maintained a good up-to-date picture of the condition of his estate, and was able to respond to building needs through the bids he entered in his major works programme.

Some use was being made of CAD systems for space utilisation exercises, but not for maintenance management purposes. The estates manager had considered, however, that there was some potential in this area, which he intended to explore in the near future.

References

(1)   Le Lucy J. (1988) Professional approach to Facilities Management, *Facilities*, *6(ii)*, November.
(2)   Powell C. (1991) Facilities Management, *CIOB Technical Information Service Paper 134*, CIOB.
(3)   Moohan, J. (1987) Profits and Maintenance Decisions of Owners Who

Rent for Income. In Spedding, A. (ed.) Building maintenance and economics – transactions of the research and development conference on the management and economics of maintenance of built assets. E. and F. Spon, London.

(4)   British Standards Institute (1984) *BS 3811: 1984 Glossary of Maintenance Management Terms in Terotechnology*. HMSO, London.

(5)   Hales, Colin (1993) *Managing Through Organisations, the Management Process, Forms of Organisation and the Work of Managers*. Routledge, London.

(6)   Shaw, Josephine (1991) *Business Administration*. Pitman, London.

(7)   Appleby, Robert C. (1987) *Modern Business Administration*. Pitman, London.

(8)   Handy, Charles B. (1993) *Understanding Organisations*. Penguin, London.

(9)   Audit Commission for Local Authorities in England and Wales (1988) *Local Authority Property – A Management Handbook*. HMSO, London.

(10)   Audit Commission for Local Authorities in England and Wales (1988) *Local Authority Property – A Management Overview*. HMSO, London.

(11)   Local Authorities Management Services and Computer Committee (1981) *Terotechnology and the Maintenance of Local Authority Buildings*. LAMSAC, London.

(12)   Porter, Ray (Watts and Partners) (1992) Letter to the editor, *Building Surveyor*. February.

(13)   Audit Commission for Local Authorities in England and Wales (1986) *Improving Council House Maintenance*. HMSO, London.

(14)   Local Authorities Management Services and Computer Committee (1981) *Terotechnology and the Maintenance of Local Authority Buildings*. LAMSAC, London.

(15)   National Federation of Housing Associations (1990) *Committee Members' Handbook*. National Federation of Housing Associations, London.

(16)   National Federation of Housing Associations (1987) *Standards for Housing Management*. National Federation of Housing Associations, London.

(17)   National Federation of Housing Associations (1992) *Housing Associations After the Act – Research Report 16*. National Federation of, Housing Associations, London.

(18)   National Federation of Housing Associations (1987) *Standards for Housing Management*. National Federation of Housing Associations, London.

(19)   Griffiths, Sir Roy (1984) *NHS Management Enquiry*. HMSO, London.

Chapter 3
The Design/Maintenance Relationship

Introduction

Within the framework of the definition of maintenance policy in BS 3811 is the term 'acceptable condition'. The condition of a building is central to the notion of building performance and must be considered throughout all phases of the building life cycle, as it is a key element in the formulation of maintenance policy.

A prerequisite for a considered approach to managing building performance is for the organisation, not only to have in its possession an accurate picture of its estate, but also for it to take an active role in its development. This begs a number of questions in relation to the manner in which the organisation procures its buildings and, in particular, to how they are delivered to it.

For example, it might well be concluded from a cursory study of many business organisations that there is little apparent relationship between many of them and the design process.

> 'Maintenance and design are frequently treated as if the two activities were unconnected ... Maintenance sections often appear to be self contained ... [leading to] ... risk of undesirable divorce from other related functions. The status of the section and its personnel may be such that there is a lack of influence at policy and strategic levels'.[1]

This view compels consideration of the design/maintenance relationship in rather wider terms than at first seems necessary. A design may be executed perfectly well within the terms of reference that have been laid down, but fail to perform properly if these parameters are imperfectly set. It should also be appreciated that, even if a perfect design is executed from an ideal brief, this design has to be developed and realised effectively. Most importantly, the process should culminate in proper delivery of the building to the owner/occupier, together with the information necessary to ensure that it is operated and maintained correctly.

This chapter is, therefore, concerned not only with design in the narrower sense, but also with the whole range of activities from inception to delivery that will ultimately influence the building's performance and the maintenance of that performance.

Performance requirements

It is a requirement in managing anything effectively that there should be a clear set of objectives. In the case of managing buildings, this includes the need to maintain the building in such a condition that it is capable of performing the function for which it was conceived[2]. Ideally, this should be represented by a stated set of criteria, against which achievement of this objective can be evaluated. However, for a substantial majority of buildings this statement is not available. The following scenarios are typical examples.

(1)   The building is old and has undergone radical changes of use during its life. If the change of use has been the result of a properly conceived refurbishment exercise it should, however, be possible to produce a definitive performance statement.
(2)   The building has been conceived in such a way that these criteria were never properly established.
(3)   The information is available but has been communicated imperfectly, or not at all, to the client or occupant.

As well as being of concern with respect to the design process, this deficiency will seriously impede attempts to manage a building effectively. Even where there is a clear statement of building performance objectives at the outset, and a perfect solution generated, the situation is a dynamic one, and proper management of the facility must recognise this and respond to it. This requires a process of on-going monitoring to facilitate effective action. Such monitoring can only be carried out properly by reference to a clearly stated performance model.

All the building systems should be included in such a model, as they will all make a contribution to the effective performance of the building's function. Of course, management judgement and policy may attach differing levels of priority to each element, but at least management will be operating within a comprehensive framework.

The process begins at the inception stage and, assuming that the need for a building (or perhaps a major refurbishment) has been identified, the next step is for the client to provide a brief of his requirements for the building. It may be presumed, initially, that a full brief development will address all the relevant issues, resulting in a design that is a proper response to the requirements of the building owner, user and contents. In ideal circumstances, following this, the design would represent a perfect model of the proposed facility which, as well as being of relevance to detailed design and construction, will be of value to the maintenance manager and the users/owner of the building, thus emphasising the need to view the building as a facility.

The design brief

Having identified the significance of brief development and assumed, perhaps rather optimistically, that this might be perfect and elicited a correspondingly ideal response from the designer, it is now necessary to examine the ways in which this may be properly developed.

Between the initial desire for space, its first occupation and continuing operation, a number of activities have to be successfully undertaken by a diverse range of people, who may only have an indirect relationship to the primary objectives of the organisation, and an imperfect understanding of them.

The effectiveness with which these activities are carried out will have a major influence on the success or otherwise of the venture. It is, therefore, essential that all parties have a common basis against which these activities can be initiated, planned, monitored, judged and controlled, and that there is a communication system that properly co-ordinates the activities of the people involved.

In seeking the evolution of a solution to satisfy the requirements for a building, four guiding principles can be identified:

- ❑ To produce a building which is appropriate and efficient for the functions it houses
- ❑ To produce a building that provides the optimum physical and psychological environment for the contents of the building, both animate and inanimate
- ❑ To produce a building which strikes an appropriate balance between initial and operating costs
- ❑ To produce a building which is consistent with the needs and aspirations of the community at large

In simple terms, the task is to seek a satisfactory design which is a solution to a set of problems posed by a series of questions. It is the framing of these questions that may be the most difficult part of the process.

Initially a building will be conceived in very general terms, with a broad statement of the building's function, e.g. factory, housing, school or office, which will then be fleshed out to produce information concerning size, spatial requirements, number of people, and facilities. This process often progresses with indecent haste and insufficient real analysis[3]. Even at this very early stage it is essential that the major components and systems of the building are clearly identified, so as to define the way in which they need to respond to the broad objectives. An organised methodology is clearly necessary and a variety of guidelines may be followed to ensure that the approach taken is a systematic one.

However, simple checklists in themselves are not enough, and it must be appreciated that this should be an analytical process, with proper attention focused on the interface between the building's operational and occupational requirements, and its design requirements in technical and organisational terms.

Although the contents of a building may be both animate and inanimate, it can be assumed that the need for the building is generated directly or indirectly by human needs. The nature of these needs will be derived by the objectives of the person or persons requiring the building, which may be social, political, economic, moral or religious, etc. The building is, therefore, required to satisfy a set of goals leading towards these longer term objectives.

Because the process is concerned with human activity, it must be supported by an understanding of the way in which humans interact with their environment. An analysis of user requirements requires consideration of the physiological, psychological and socio-political-economic environment. Human beings are not only shaped by this environment but impinge on it; in other words, they are not passive, and this may complicate the problem.

The provision of any building must also satisfy objectives which are common to a range of potential users, and some which are specific to individuals or user groups, all within an organisational entity (figure 3.1). To create an optimum solution therefore requires the resolution of conflict, and necessitates some compromise. For example, the objective of survival is fundamental. The goals that each individual may see as a means to achieving it will, however, differ as other objectives may interact with this one.

Each person will have a set of these goals but no two people will normally have quite the same set (figure 3.2). In the case of buildings, it is necessary to consider not only individuals but groups of individuals that

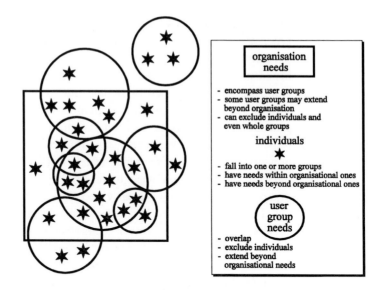

Figure 3.1 Individual *v.* group *v.* organisational needs.

	user needs to be satisfied					
	1	2	3	4	5	6
user one	★	★	★			★
user two		★	★	★		★
user three	★	★	★	★	★	★
user four	★	★	★	★		
user five		★		★		
user six		★	★	★		
user seven	★	★	★			
user eight			★		★	
	4/8	7/8	7/8	5/8	2/8	3/8

Figure 3.2 Simple ranking analysis of user needs.

exist as a corporate entity of one form or another. Their objectives and goals will be the product of a complex process which will have taken place in the wider 'universe' (figure 3.3).

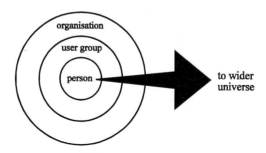

Figure 3.3 Human needs inter-relationship.

To determine the requirements of the future user of a building it is essential to understand these goals and objectives as fully as possible, and this must be the starting point for the execution of a user requirement study. The level of understanding developed through this study will inform later studies, and be instrumental in the success or otherwise of the brief.

The next step is to translate what may often appear to be rather abstract ideas into a set of more concrete criteria. The more effectively the client is understood by the designer, then the more accurately this task may be performed.

A building performance model

The Building Performance Research Unit at the University of Strathclyde have produced a conceptual model of the building performance system and people[4] which is summarised in figure 3.4. This represents an idealised view of building performance and, despite the passage of time, still provides a working model which may be of value in considering the nature of a design brief.

Using this model, the brief can be seen as the analysis of the objective system, through which broad goals may be set. This requires reference to the other systems, to help shape the questions that have to be asked in order to address these goals. Thus, a question and answer routine emerges that facilitates a move from the general to the particular (figure 3.5). For example, the process may start from a broad goal that requires the provision of essential climatic modification. This general require-

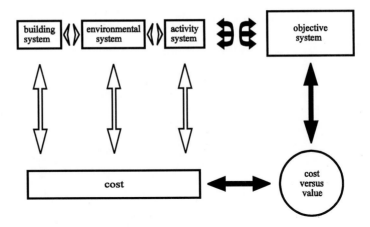

Figure 3.4 Building Performance Model (*after* Strathclyde University BPRU).

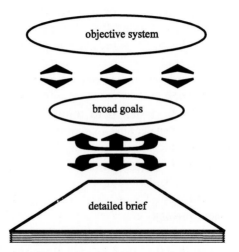

Figure 3.5 From objective system to brief.

ment may need to be qualified in view of the need to match other aspirations of the client. He may, for instance, have a specific need to meet a certain standard of acoustic performance because of the nature of the business. This may be satisfied by manipulation of either the building system, the environmental system or the activity system, or a combination of these (figure 3.6).

Another climatic modification goal, however, may be a requirement for a building with a rapid thermal response, therefore suggesting the use of a lightweight building system. This will militate against the satisfying of the acoustic requirement through the choice of building fabric, and

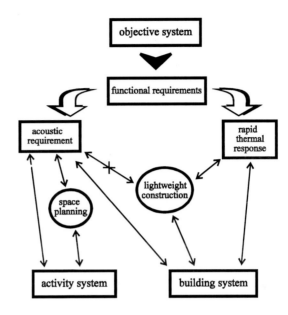

Figure 3.6 Interactive analysis based on the Building Performance Model.

suggest the need for an alternative solution to this objective, perhaps through an analysis of the activity system.

It is clear that the formulation of a brief through a user requirement study requires the posing of a series of hierarchically structured questions, and that this is an iterative process. The objective of such an approach is to analyse the total building problem, and a properly considered design decision cannot evolve without it. Asking the right questions is only part of the solution, but can increase the expectation of eliciting the information necessary to produce an accurate brief.

A number of diagrammatic techniques are available, but these are essentially informative ways of presenting the problem, rather than providing solutions (figure 3.7). In the past some use has also been made of optimisation techniques from operational research. However, the problem is fundamentally a behavioural one.

Format of the design brief

Although the analytical process outlined above is essentially an intellectual activity, there is a fundamental requirement for a methodically structured representation of the brief. This will be necessary, not only for design purposes in order to provide a clear statement of the requirements of the building, but it will also be useful later for all those concerned with

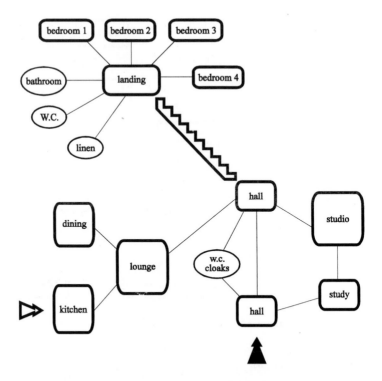

Figure 3.7 Interaction diagram for design of designer's house and studio.

the effective management of the built asset. The format outlined below is one approach that may be taken.

1.0   Introduction
  ❏ Scope of document and summary
  ❏ Contents

2.0   Requirements
  ❏ General
    ○ geography and location
    ○ broad function(s)
    ○ approximate size
    ○ time scale constraints
  ❏ Occupation density and spatial requirements
  ❏ Detailed functions, the relationships between them and hence circulation patterns
  ❏ External requirements, e.g. car parking and landscaping
  ❏ Communication systems perhaps related to circulation and functional relationships

    ❏ Ancillary functions, e.g. catering, cleaning
    ❏ Special requirements identified immediately, e.g. air conditioning and IT

3.0   Overall project objective(s)

4.0   Design criteria
    ❏ Space and environment
    ❏ Site works
    ❏ Building shell
    ❏ Interiors and finishes
    ❏ Electrical services
    ❏ Heating, ventilating and air conditioning
    ❏ Plumbing
    ❏ Process and/or production systems
    ❏ Communication systems
    ❏ Fire protection and means of escape
    ❏ Transportation and circulation systems

5.0   Cost constraints/objectives

6.0   Management of the project

The design criteria identified will need to be supplemented by more detailed information, perhaps in the form of a design criteria sheet. It is recommended that these detailed analyses are included in an appendix, cross-referenced to the above items. The design criteria sheets set out the performance requirements of the space, and the activities to be accommodated within the building. A simple illustrative example is shown in figure 3.8.

Maintenance and the brief

Determining the requirement of a building from a study of the user will set in train a decision making process which will inescapably impinge on the maintenance requirements of the building. In the first instance there may be a direct effect, in that the broad goals of the building may include a particular stance by the user with respect to running costs. Subsets of goals may then be developed to cover issues such as the minimisation of energy usage and low maintenance characteristics in the building fabric. In such cases, it follows that these items would be overtly identified in a brief statement, such as that outlined above. However, even if low maintenance is not directly on the agenda, a less than accurate definition of the requirements of the building may lead to inappropriate planning

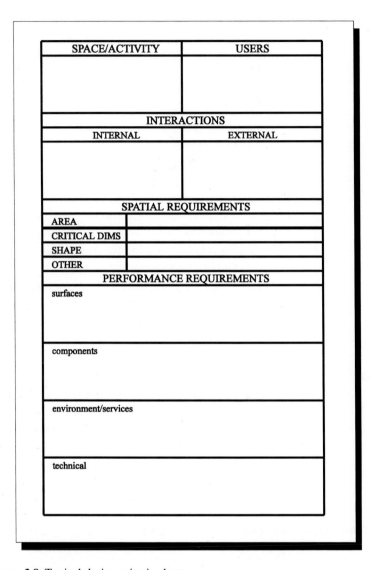

Figure 3.8 Typical design criteria sheet.

and choice of components, materials, etc., any of which will lead to poor
building performance, with the implications this has for its maintenance.

As long ago as 1978, a NEDO report[5] concerned with client perceptions, concluded that:

'According to a number of companies, the standard of service given by
the building industry relates closely to the amount of effort expended
by the client in establishing a good brief at the beginning.'

Brief execution

It is important that, at subsequent stages of procurement, proper consideration is given to the information assembled at the briefing stage, and, furthermore, that it is effectively communicated.

From the design team's perspective, it is usually considered desirable that the individual who will be responsible for the design should be the one to collect the brief from the client. However, effective brief collection and management is a highly skilled business, and not all designers, even the most talented, possess the necessary attributes to do it well. It is important, therefore, that the right person is selected to undertake this task, irrespective of who he or she is.

Many large clients now realise the importance of properly developing a brief within the context of their organisation's entire business and financial plan/strategy. In order to ensure that the brief accurately reflects their needs, they increasingly engage the services of a 'professional client' or appoint a project manager. This individual may either be an 'in-house professional', or an external consultant.

In recent years there has been a rather more flexible approach to building procurement leading to an increased use of management contracting, and design and build methods. Whilst these procurement methods provide opportunities, there is some cause for concern as impetus for their development has tended to come from clients, principally, with the objective of speeding up the procurement process, rather than as a means of securing an improvement in the performance of the building. Moves towards extensive 'fast-tracking' of the procurement process appear to place increased pressure on an already strained system, particularly in respect of preparing detailed designs.

There is also a feeling that, no matter which procurement procedure is adopted, increasing competitive pressures result in cost-cutting being focused on those areas that are perceived as making a less productive contribution. Brief development is one such activity that tends to be marginalised. Many building professionals express some concern that this scenario will leave as big a legacy in the next century as the mistakes made in the 1960s.

Design and construction

Outline proposals and sketch schemes

The next stage in the design process is the production of a sketch scheme. This will, in all probability, overlap with the development of the brief, and the process is very much an iterative one[6].

All too often a fracture occurs in the logical development of the design, when the concept used for the building is a product of the designer's objectives, rather than those of the client[7]. If proper regard is not given to an accurate interpretation of its findings, the briefing exercise will have been of little value. This strengthens the case for a separate person to carry out the user requirement study and write the brief, even though this may be at the expense of more direct lines of communication.

Following the formulation of a sketch scheme comes the important stage at which the designer seeks the client's final approval of the proposals. As the client's perception of the scheme often represents a non-professional interpretation of what is proposed, it is essential for what is on offer to be objectively evaluated against what is required. Here the client will benefit substantially from the advice of an in-house professional or project manager. If accurate judgements are not made at this point then the project may be launched onto a course that is misdirected.

Consideration of building performance issues are rarely fully evaluated at this stage. Catt[8], for example, has pointed out in a series of articles on planned maintenance, the importance of undertaking an early appraisal of design from the maintenance aspect.

Research at Heriot-Watt University[9] indicated that the interest of the design team in maintenance varied over the life of a project, with concern increasing towards the latter stages, where it was largely influenced by economic and practical factors. This confirms a general fear of an irregular and inconsistent approach to the issue.

Detailed design and building defects

It is during the development of the detailed design that there is the greatest perceived scope for technical performance problems to originate. There is a general assumption that bad detailed design practices are the source of a large proportion of performance problems during the life of a building. However, it could be argued that in many cases the real cause lies at an earlier stage.

Defects that occur during the lifetime of a building are all too easily attributed to detailed design, and this may be a hindrance in fairly allocating the blame. Failures, deemed to be the result of fair wear and tear by the expert, may be seen in a completely different way by the building owner. The view taken by the latter must depend upon the original performance expectation. There is thus not only a duty on the designer to determine what these performance requirements are, but also a need to make clear to the client what is a reasonable level of performance for a given material or component. Neglecting to do this may lead to many deficiencies that should be classed as defects to be accepted as normal maintenance work, or their rectification as an improvement. It may also lead to the development of a perception amongst building owners, about the nature of their buildings, that is decidedly unhelpful to a maintenance team.

Normal maintenance, in its strictest sense, ought to consist solely of those actions needed to maintain an element or component in the condition required to perform the function for which it is designed. However, there is evidence to suggest that around 20% of so-called maintenance expenditure can be identified as being necessitated by defects. In general, these tend to manifest themselves early in the life of the building and, whilst the majority are associated with structural matters, there are significant incidences of poor performance of finishes, difficult access problems to services and other detailing problems. Action to remedy many of these may also be classified as improvements, e.g. creating additional access to service ducts, whereas in reality a substantial proportion should be considered to be design defects.

During the early 1970s the Building Research Advisory Service made detailed studies of samples of buildings and their analyses reveal that the largest single type of fault resulted from making wrong choices about materials or components for a particular situation in the building. The poor decision making leading to this may have occurred at either the detailed design stage or in the user analysis.

Lee[10] considered that maintenance determinants could be normal or abnormal, when classified on past experiences and current expectations, and each of the following were influential:

❏ The adequacy of the design and the suitability of the materials specified
❏ The standard of workmanship in the initial construction and subsequent maintenance operations
❏ The extent to which the designer has allowed for present and anticipated future needs

Detailed design decision making

A major issue, during detailed design, is the selection of materials and components, and choice is becoming increasingly difficult. There are immense pressures tending to change the nature of building from a traditional craft process towards an assembly of factory-made components. This has several consequences. In the first instance, the designer is often faced with having to choose between an extremely large range of products, all of which on face value may satisfy performance requirements. Additionally, designers are subjected to a great deal of lobbying to use one product rather than another, and are bombarded with a plethora of publicity material.

This can have two extreme consequences:

- ❑ It may sway the designer into using a new component which has not been sufficiently proved in practice, with a consequential failure to meet requirements.
- ❑ On the other hand there may be the opposite reaction, resulting in the designer using a component or material with which he is familiar, whether or not it is the right choice in the circumstances.

In April 1975 the BRE[11] concluded a research project by commenting that the major short-coming in detailed design appeared to be a failure to make use of the authoritative guidance that was available. This might be a simple failure to use codes and standards properly, or attributable to the shear volume of advice. In many cases, they suggested, there was what amounted to a perverse avoidance of the standard solution, on the basis that the designer could develop detailed design from first principles. '... the emphasis is on prima donnas who think out each new problem afresh.'[12]

There is also evidence suggesting that much detailing may be excessively complex, accompanied by over ambitious specifications, and that the designers' argument that this is in the interests of improved quality is not proven[13].

The changing nature of building construction, to make greater use of component technology, places much greater emphasis on the need to co-ordinate design from both the technical and organisational viewpoint. From the technical viewpoint, this manifests itself in a steadily rising incidence of joint failures, which is undoubtedly due to a poor understanding amongst designers of the performance limitations of joints and sealants. This ignorance is mirrored in substandard execution on site and inadequate supervision. All too often it is clear that the designer asks too much of a detail and, furthermore, does not understand the nature and limitations of site operations.

From the organisational perspective, the move towards 'component-based' construction exacerbates a problem that has existed for some time, in that the design and construction of a building is not only about the work of one designer, but of a team of people working in different organisations. Current trends suggest that this fragmentation is increasing.

The responsibility of the design/construction team

Specialist designers are invariably employed on all but the smallest projects and the task of co-ordinating their efforts, whilst by no means being easy, is all too often given insufficient attention. The reasons for this are diverse and cannot be fairly attributed to any one body. There have been a number of views expressed over the years, the most common of which point the finger at the alleged shortcomings of the architectural profession in terms of its management competence. However, at a seminar organised jointly by the RIBA, the Association of Consulting Scientists and the Council for Science and Society in 1977[14] the following observations were made.

(1) Improving design competence can make the biggest contribution
(2) Execution on site is the next most important item and faulty materials/products are a comparatively minor problem
(3) Lack of design competence is rooted fundamentally in an education system run largely by theoreticians rather than practitioners
(4) At the level of design procedures, the greatest need is for improved checking systems
(5) Site supervision and the checking of materials and workmanship must be improved
(6) Possibly the greatest problem faced by the designer is design proliferation
(7) There is too much unnecessary innovation and not enough reliance on standard, tested solutions of known reliability

This seems a balanced view, and provides a perception of a problem that is a function of the whole process, and where blame must be shared.

The client's attitude should also be questioned, and Bowyer has argued[15] that the two principal factors to be considered in the initial stages of a construction project are time and money. He claims that there is rarely time to execute the preliminary stages, or the detailed design work, in the most effective way and that this situation stems from a client's ignorance of the information needs and procedures that must be undergone. Additionally, the financing of buildings, both in the public and private sectors, is subjected to pressures that do not encourage proper consideration of the building's long term performance. Consequently he concludes that:

'... the Architect is caught between the two forces of limited time for his professional work and the problem of designing a sound and suitable building within unrealistic cost limits. The Architect who demands proper time for his work and provides realistic costs for the project rarely survives in a world dominated by the accountant.'

The building contractor must also shoulder some share of the blame, be prepared to make greater efforts to understand new technology and allocate proper resources to site supervision. Contemporary developments in procurement procedures, whilst often at the client's behest, are not improving this situation, and the widening of responsibility does not help accountability.

There have been a succession of pleas calling for greater involvement of the maintenance manager at the design stage, and also for a more carefully considered view of the whole building procurement process which recognises building maintenance as an important integral part of it. A particular criticism, in this respect, has been of the procedures adopted for handing over, commissioning and running in of the building, and a failure to provide good quality feedback information. All of these aspects have been the subject of frequent recommendations over the years, but infrequent action by the industry, representing a major failure in terms of the service it provides to its clients.

The Construction (Design and Management) Regulations 1994

Health and safety on construction sites has been an important issue for some time with the major responsibility lying with the contractor. The Construction (Design and Management) Regulations 1994 (CDM)[16] came into force on 31 March 1995, and their implications are likely to have an important influence on both the execution and management of maintenance. This is because, in principle, they represent a major extension of responsibility for safety from the contractor to the designer and the building owner.

Determining when the regulations apply is rather complicated, but in simple terms, they are applicable to all but the smallest jobs. These are defined as jobs that:

- Will not be longer than 30 days
- Involve no more than 500 man-hours of work
- Do not employ more than four people at any one time

In-house operations do not come within the scope of CDM. Work to a person's own house is also exempt, provided it is not part of a trade or business. However, demolition works are always subject to CDM, unless for a householder or in-house. CDM is always applicable to design work, no matter how small the job.

There are also complicated rules to determine what constitutes a job or project. Under the regulations the project ends with hand-over and occupation. Operations subsequent to this constitute a new project. Determining the applicability of CDM to maintenance operations is likely to be a complicated business.

The European Directive on Temporary or Mobile Work Sites[17] defines the term 'project supervisor' and under the UK CDM regulations this role is divided into two, with responsibility resting with 'the planning supervisor' and the 'principal contractor'. There was an early recognition that no one professional was necessarily equipped to fulfil the role of planning supervisor and this has given rise to the creation of a specialist consultant. The planning supervisor is appointed by the client to:

- Notify the Health and Safety Executive of the existence of the project
- Ensure that designers fulfil their responsibilities under CDM
- Ensure that designers co-operate on site safety matters
- Ensure that a health and safety plan is produced
- Ensure that a health and safety file is produced and amended as necessary and handed to the client on completion
- Give advice on appointments

The designer's responsibilities are to:

- Inform the client of his responsibilities
- Avoid or reduce foreseeable risks on site
- Ensure that the design has adequate information on health and safety
- Co-operate fully with the planning supervisor

The approved code of practice[18] gives a great deal of advice with respect to the interpretation of these duties. Furthermore, there is an acceptance that health and safety has to be measured in the context of overall building design and construction requirements and, in particular, economic realities. A cornerstone of the requirements is the term 'risk assessment' and this is likely to be a contentious issue, particularly when making judgements with respect to alternative designs solutions. There will undoubtedly be a great deal of confusion as to the strict interpretation of these responsibilities, which will only be clarified after a body of case law has been developed.

Under the designer's duty to produce designs that can be safely built, the most important consideration will be access, not only during construction but also for cleaning, servicing, maintenance and repair. The client must automatically receive a health and safety file at hand-over, warning of any potential risks, and the designer is responsible for providing the planning supervisor with the information to be included.

Whenever the CDM regulations apply there must be a client and this is defined in the regulations as 'any person for whom a project is carried out'. There are no requirements in respect of client competence, nor for adequate resourcing (unlike the appointment of designers, safety supervisors, principal contractors and other contractors). The client's responsibilities are to:

❑ Appoint a planning supervisor and a principal contractor
❑ Ensure the contractor is a contractor under the definition of the regulations
❑ Ensure that the planning supervisor is competent under the provisions of the regulations
❑ Ensure that the appointees are adequately resourced
❑ Ensure that the health and safety plan exists before the work commences
❑ Ensure that the planning supervisor has information about the condition of the premises
❑ Ensure that the health and safety file is available
❑ Pass on the health and safety file to anyone acquiring an interest in the property

Clients can delegate all or any of these duties to an agent, and it must be borne in mind the major role of the planning supervisor is to advise the client. Of major importance is the requirement to provide the planning supervisor with information relevant to his duties, and this imposes a major obligation to maintain accurate and up-to-date building records which will include:

❑ Previous safety files
❑ Condition survey information
❑ Maintenance arrangements
❑ Details of plant and equipment
❑ Occupational information, both past and present
❑ Covenants and tenancy information

Life cycle costing[19–21]

Building owners, as a rule, place undue emphasis on the capital costs of a project at the expense of future running costs. If a proper balance is to be struck, this issue needs to be addressed whilst the brief and the detailed design are being developed. Various techniques have been developed to assist in the analysis of a building's total costs over its life span.

The life cycle cost (LCC) of an asset is defined as the present value of the total cost of that asset over its operating life, including initial capital cost, occupation costs, operating costs and, the cost or benefit deriving from disposal of the asset at the end of its life.

The broad objectives of a life cycle costing exercise may be:

❑ To enable investment decisions to be made more effectively, taking into account all costs that may arise from it
❑ To consider the impact of all costs, rather than just capital costs
❑ To provide information that can contribute to the more effective management of the completed building
❑ In the context of building procurement, to assist in the evaluation of alternative solutions to specific design problems

For example, the technique may be applied to an element, such as a flat roof, where there is a need to make a decision between a high initial cost, low maintenance, long life solution and a low capital cost, high maintenance, short life one. This may be carried out using discounting techniques to compare present and future costs for each alternative on a common basis.

The important point to note about the use of such a basic example as this is that its evaluation is far from simple. Indeed, such evaluations are complex, and a proper analysis requires consideration not only of direct costs and/or benefits, but also indirect ones. Some of these may be readily quantifiable, such as disruption costs due to failure and/or repair execution, whilst others will require more subtle forms of evaluation. An example of the latter may be an aesthetic consideration in a sensitive part of a building, such as the reception area of an important office facility or hotel.

Sound judgement is needed when calculating life cycle costs, together with the exercise of technical, managerial and financial skills.

Implicit in the term life cycle is the notion that, during its life, a building progresses through a number of phases. The nature of the cycle will be dynamic, as opportunities for change, renewal and adaptation present themselves, and this must be recognised when calculations are being made.

The sequence of the life cycle phases is described, in engineering terms, in BS 3811 and for the life of a building this is:

❑ Brief collection and development
❑ Design
❑ Construction
❑ Commissioning

❑ Maintenance
❑ Modification
❑ Replacement

An essential element in using LCC is defining the life cycle period, or 'building life', and within this, component and/or element lives.

Life cycles, obsolescence and utility value

Buildings are, in general terms, very durable and, if properly maintained, may last for centuries. Even if maintenance is indifferent, a physical life of 50–60 years is quite realistic. The term physical life here means the life of the identity of the building, and not necessarily the life of every physical part of it. The attitude of the client to the required physical life of the building may be established in preparing the brief, although it might not be the subject of a specific statement.

Many components will need replacement during the life cycle, some several times. The length of time for which it will be worthwhile to continue to repair and renew parts of a building will depend on how well the building continues to meet the needs of the function for which it was built. In economic terms, past costs are of no importance, and only the relationship between future costs and future value is significant. A building is worth repairing if the future utility to be derived from it exceeds that which could be obtained by demolishing it and erecting a new one.

In an extreme case, it can be worthwhile demolishing a new building when, for example, the purpose for which it was built has ceased to exist. It is worth noting that this situation might arise as a result of a bad initial investment decision by the client, an inadequately managed inception process, or perhaps a poor response by the professional advisors. At this point, of course, the possibility of adaptation may be considered, and in a sense this could be said to give the building a new life, in that its identity has been changed.

Any consideration of building life must take account of the notion of obsolescence, which relates to economic considerations, directly or indirectly. The RICS[22] has identified the following six forms of obsolescence:

❑ Economical
❑ Physical
❑ Functional
❑ Technological

❏ Social
❏ Legal

Economical and functional obsolescence are probably the most common of these. However, all these forms are interlinked, and the simpler categorisation of Ashworth[23] into physical, functional and economic lives may be more realistic.

As obsolescence is concerned with the potential utility that may be derived from a building, some thought must be given to the rather contentious issue of value. In the commercial sector this may be something which is clearly objective, and value can be stated in money terms, with the value of the building being determined by its worth as a factor of production. In the service sector the notion of value is much more difficult to isolate, and it is often here that building fabric performance is of central importance. Value, therefore, is determined by not one but many attributes.

The assessment of a building life is a difficult exercise, requiring good judgement, knowledge of context and previous records. For business accounting purposes, the building life is the period over which the relevant organisation holds an interest in the building. At the end of this period the building may have a residual value and pass into the interest of another organisation. The application of LCC must always be carried out in a proper context, as considerations of a life cycle may differ somewhat from notions of the life of a building or a component as normally perceived.

Component life may conceptually be rather easier to define, if not to evaluate, as it will generally be shorter and more easily foreseeable than for the building itself. The circumstances leading to the replacement of a component may include deterioration, failure or obsolescence, and the latter should be considered separately from the others. The issue of component life is a very important one for its own sake, and not as simply part of a life cycle costing exercise.

Life cycle costs

Life cycle costs can be divided broadly into three categories:

(1) Capital costs which, in theory, may be relatively easy to predict
(2) Costs-in-use which are more difficult to predict, and include several components
(3) The costs involved in final clearance of the site or disposal of the asset

Against these costs can be offset the residual value of the asset, i.e. financial inflows accruing from disposal. A simple view therefore is:

Life cycle cost = Capital cost + cost-in-use − Net residual value

Maintenance costs will only be part of the cost-in-use, and all recurring costs must be fully considered. Generally the three Rs of running, repairs and replacement will cover costs-in-use and these may all be subject to consideration during the design process.

Exhaustive lists are difficult, but the following will typically need to be counted:

- Maintenance, including redecoration
- Energy consumption
- Cleaning
- Rates
- Insurance
- Estate management or management overheads
- Finance costs

There is also some debate as to whether the costs of alterations and adaptations should be considered and, whilst this is a difficult issue in definition terms, there is no doubt they must be part of the life cycle cost equation, irrespective of whether or not they were predicted. For practical and operational purposes, and the factors discussed above, it is not possible or indeed realistic to completely ignore the question of defects, repair/replace decisions, and improvements that will be necessary to maintain the utility of the building. Expenditure made to materially change the function of the building, i.e. change its identity, will however be relegated in significance.

Financial appraisal techniques

The analysis of costs and benefits

For accounting purposes, it is convenient to divide costs into capital and revenue. Expenditure in each of these categories is often determined by different decision making processes, and this is a major obstacle to the efficient implementation of LCC. The main life cycle costs are summarised in figure 3.9[24].

One of the purposes of carrying out a LCC exercise is to assist a decision making process during design. This may be at the global level, when strategic design decisions are being made, as well as at the detailed

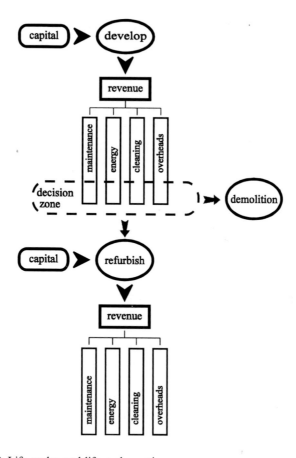

Figure 3.9 Life cycles and life cycle costing.

design stage, when it is necessary to decide between the relative merits of different specifications, and where relative life cycle costs will help inform the choice.

There are many techniques of financial appraisal that can be used, the simplest of which is the payback period approach. This is the time taken for the returns from an investment to equal the initial outlay, and is sometimes used to evaluate the financial benefits of a variety of energy conserving measures. For example, if roof insulation costs £200, and it is estimated that the measure will reduce fuel bills by £40 per annum, the payback period is 5 years. The major weakness of this method is that it does not take into account the time value of money.

LCC techniques are concerned with the evaluation of a stream of costs over time and with present value. This implies that to evaluate costs properly there must be a common basis of comparison, and future costs need to be converted to present day equivalents by discounting them. Appendix 2 explains the principles involved in the discounting of future costs, allowing them to be compared with expenditure now, on a common basis.

The illustrative example given evaluates the relative merits of two alternatives, A and B. The former involves a capital outlay of £100 000, with a predicted annual maintenance of £2000. The alternative has a capital cost of £120 000, with annual maintenance expected to be £500 per annum. If the time value of money is ignored, after 60 years alternative A involves a total outlay of £220 000 and B £150 000.

If the calculation is repeated, taking into account that £1 in the future is worth less than £1 today, then the picture is not so simple. In the example given in Appendix 2, a discount rate of 9% over a period of 60 years suggests that a comparison on a present day value basis makes A the cheaper option.

The calculations can be performed in three main ways, but all based on the same principle.

(1)   The simple example referred to above discounts all future cash flows to give a **net present value**.
(2)   Alternatively the cash flows over the life of a building or component can be converted using discount tables to an **annual equivalent cost**.
(3)   The third method that can be used is to determine the interest rate, which when applied to all future cash flows, gives a net present value of zero. This is then called the **internal rate of return**.

For purely comparative studies, either of the first two methods may be used, but the use of an annual equivalent may be of additional benefit for financial management purposes. The internal rate of return may be a useful measure in commercial decision making, but is of less relevance for present purposes.

In terms of the design decision making process, there are several characteristics that need to be examined with respect to the discounting techniques.

(1)   The techniques assume that the designer is choosing from mutually exclusive options. In practice they may have a larger range of options available that are not mutually inclusive. Design decisions are rarely a simple either/or choice.
(2)   Only negative cash flows are being considered, and the positive benefits of each, whether or not they can be evaluated in cash terms, are not included. Any comparison purely on these grounds is assuming that each option gives the same utility. Public sector

organisations, for example, have many demands to satisfy, which will include social and legislative factors. In such cases the use of cost benefit analysis may be desirable[25]. In the private sector, too, not all benefits and costs can be evaluated in pure cash terms, although it may be more realistic to attempt to measure utility in a theoretical manner.

(3) It should also be pointed out that a fully rigorous life cycle analysis should count all costs, including those that stem indirectly from maintenance operations, such as the cost of disruption to the every day performance of the building.

(4) The use of the techniques assumes the cost of maintenance expenditure is determined on the basis of predicted needs and, given the inevitable pressures on maintenance funding, these might not be fully met. The techniques described might show that a low cost, high maintenance solution is preferable under some circumstances. However, the question has to be asked whether in reality funding for the maintenance work will be forthcoming when resources are scarce. The consequent deterioration of a building or element, if not properly maintained, may severely impair the performance of the building to such an extent as to render the original analysis fatuous.

Inflation

So far no account has been taken of inflation and there are several reasons for ignoring this, including the difficulty of predicting it over a building's life. For comparative studies, it is assumed that inflation will affect each option in the same way and can therefore be ignored. This is rather questionable, as the error attached to proceeding on this assumption may well be greater than has hitherto been assumed. For example, under some market conditions, labour costs may rise at a greater rate than material costs and, as maintenance tends to be more labour intensive than new-build, substantial errors may accrue.

In the final analysis, however, it is clear that any attempt to allow for inflation will greatly complicate the process and, given the range of uncertainty attached to all the input data, it is probably wise to avoid too sophisticated an approach.

Uncertainty and data quality

All the techniques require the forecasting of future cash flows and are hence subject to substantial uncertainty. It has been argued that, for comparative purposes, this may only have a limited influence. It is also true that the greatest uncertainty attaches to long term cash flows, and the examples in Appendix 2 show that these will be subjected to very small discount factors, so that the margin of error is correspondingly smaller.

What is certain, however, is that data on the life of a building and its components is an essential pre-requisite. Such data is difficult and expensive to obtain and to update, particularly if it is to have statistical significance. A summary of appropriate statistical techniques that may be useful in assessing maintenance data is given in Appendix 1.

Economic life

All discounting techniques require an assumption to be made about the life span of the building under consideration. Traditionally, a 60 year life has been taken for analyses such as those discussed above. The points made earlier with respect to notions of economic life and obsolescence must, however, be borne in mind. One retail chain, for example, takes an economic life to be 11 years because, it is assumed that, at the end of this period the market will either have grown or disappeared, thus requiring either the disposal or replacement of the building. This is becoming increasingly true of many modern industrial buildings.

The notion of the disposable building must be more readily accepted than hitherto and investment decisions adjusted accordingly. In such a scenario it is reasonable to expect this requirement to be an important aspect of user requirements, which should be clearly identified during the collection of the brief, and responded to accordingly.

Interest rates

LCC techniques also require the prediction of interest rates, and this is an area of undoubted concern. The higher the interest rate the lower the present value of future costs. In the comparison example given earlier, a 9% discount rate rendered option A, with a lower capital outlay, the cheaper. At a 6% discount rate, option B becomes the cheaper option in present value terms. It can be seen, therefore, that high interest rates favour lower capital outlay, as future maintenance costs appear less and less significant (figure 3.10).

This is a serious weakness because, as has been seen, the analysis tends to ignore many unquantifiable factors which may seriously affect the building's performance. Not least of these may be the energy implications in material usage of high maintenance buildings, not normally taken into account in discounting techniques due to a lack of usable data.

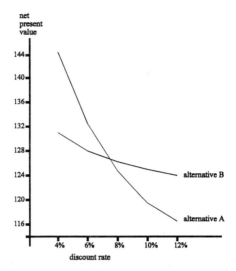

Figure 3.10 Effect of variations in discount rate.

The effect is even more pronounced when inflation is taken into account, as maintenance and running costs may be set against taxable profits, which may to some extent mask inefficiencies in the asset's performance. More sophisticated analysis of the effect of interest rates can be carried out using sensitivity analysis[26].

If a 15 year life and a 10% interest rate are taken as the base case, and a vertical and horizontal line drawn through this point, some measure of the sensitivity of the analysis to changes in the variables can be obtained. For a given interest rate it can be seen that, as would be expected, a reduction of asset life causes a steep increase in annual costs. Similarly, an increase in life causes a reduction in annual cost, but at a reducing rate, i.e. it tends to be less sensitive.

For a given life it is also easy to examine how annual costs vary with interest rates. Note that, at a 10 year life, an increase in interest rates from 10% to 20% increases annual costs by approximately £850. However, at a 6 year life, the same rise in interest rates increases annual costs by £750.

Flanagan et al.[27] have also described methods by which risk management techniques can be used to enhance the process, and improve the information given to the decision maker.

Application of techniques

The basic concept is that design decisions should take into account long term costs, as well as initial costs, and that the overall design solution should strike the right balance between these. The technique can be used at all stages in the process.

At inception stage it may be used as a tool to evaluate a range of strategies for satisfying the demand for building space. For example, in deciding between refurbishment of an existing facilitity or the construction of a new one. At the briefing stage the technique can be used as part of the two way dialogue between designer and client, in order to assist the client in this difficult decision making process. Indeed, determination of the relationship between initial expenditure and running costs, at least in principle, is an essential component of the brief. If the briefing stage has resulted in providing a proper framework for design, then LCC is an important tool throughout the whole of the detailed design phase. Furthermore, during building occupation, LCC can also be used to aid decision making concerning the framing of maintenance and renewal policies, and as part of the overall system of financial management and control.

However, these techniques should never be considered as a panacea for the resolution of all difficult decisions. They are simply aids which, together with others, can be used to supplement skill and judgement.

Value engineering

This is a related field of some interest that embraces the whole process of a building's procurement and management, and which may use some of the techniques discussed. Zimmerman[28] defines value engineering as:

'... a value study of a project or product that is being developed. It analyses the cost of the project as it is being designed.'

Value analysis on the other hand is:

'... a value study of a project or product that is already built or designed and analyses the product to see if it can be improved.'

The construction process

There has been an on-going commentary, over a number of years, concerning the standards of supervision and, hence, quality levels in the buildings produced by the UK contracting industry. Therefore, the contracting side of the industry should be subjected to the same critical scrutiny as the design side.

The move towards buildings becoming an assembly of components not only has repercussions for designers, but also places a series of different demands, both technical and organisational, on contract management. In the technical sense, the issue of joint performance is an obvious one, and the requirements this imposes on assembly, accuracy and tolerances are well understood in principal.

At the more specific level of detailing, the ability of the designers, through their knowledge and experience, to provide technical information which is both functionally correct and practical is of prime significance. What is most important, however, is the need to recognise that proper solutions to these problems do not rest with either designer or contractor alone, but require each party to have a proper understanding of each other's problems and objectives.

The organisational consequences of the changing nature of contracting impose a much broader set of problems, and these stem from two factors:

(1) The changing technology of buildings increases the number of parties to the contract. For example, the use of particular components may introduce an additional specialist at the design stage and an additional specialist contractor during construction.
(2) The motivating force behind current practices in contractual arrangements have not been driven by building performance objectives, but rather by the desire to reduce procurement time.

Whilst there may be some hidden benefits in more flexible contractual relationships in terms of performance, it is undoubtedly true that the role of the main contractor has diminished to the point where his role, more often than not, is essentially that of a co-ordinator of a series of work packages. Management contracting formalises this situation, but even on contracts let by a traditional competitive tender, the main contractor employs few, if any, of his own operatives on site. All the work is likely to be subcontracted, either to domestic or nominated sub-contractors.

Information management

The changing nature of the construction industry places an increasing premium on information management, and this was recognised by the Co-ordinating Committee for Project Information (CCPI), established in 1979 by the RIBA, RICS, BEC and ACE[29]. The object of this organisation was to address the increasing problems of managing construction information.

Despite the work of the CCPI, the quality of information management still remains a grave cause for concern and the essential link between design and construction needs to be recognised, because readily available, reliable and easy to understand project information is critical to the effective pricing, planning and control of construction work.

The following deficiencies in project information were identified by the CCPI:

- ❑ Missing
- ❑ Late
- ❑ Incorrect
- ❑ Insufficient detail
- ❑ Impractical
- ❑ Inappropriate
- ❑ Unclear
- ❑ Not firm
- ❑ Poorly arranged
- ❑ Unco-ordinated
- ❑ Conflicting

BRE studies of quality levels on building sites, between 1978 and 1983, found that the predominant causes of quality problems on site were poor workmanship and unclear or missing project information. Their research also suggested that a proactive attitude by the contractor towards information flow led to higher levels of quality being achieved on the project[30].

This study, and a later one produced by the BRE for the NEDC in 1987[31], confirmed the view that the worst causes of disruption on building sites are due to late and incomplete documentation, and that this is almost accepted as a way of life. This has serious repercussions for quality.

In the first instance, construction information produced hurriedly under pressure is more likely to be of a poor standard. Secondly, there is massive disruption to site processes, with inevitable consequences for quality. The site manager is frequently being deflected from his role of site supervisor by having to spend a disproportionate amount of time trying to rectify information deficiencies.

The problem, however, is not only one of information deficiency, but one of poor co-ordination between sets of information. This will also lead to an enormous amount of time-wasting on site in carrying out detective work to elicit the answer to a technical query, and/or to resolve conflicting information. Information conflicts themselves are a manifestation of poor co-ordination between members of the design team.

Viewed from the design office, however, there are a number of issues that contribute towards these problems. These tend to relate to the time and financial pressures discussed above. What is noteworthy is that a 1982 research paper[32], as well as attributing many of the problems to these issues, also allocated much of the blame to poor briefing.

The CCPI concluded that amongst a number of construction industry problems strongly influenced by information problems were:

❑ The incidence of technical defects in finished buildings
❑ The frequent poor quality of finished work, related to high maintenance costs

In 1987 the CCPI published the following documents, with the objective of assisting the co-ordination of project information.

❑ *Common Arrangement of Work Sections*, which provides an efficient and acceptable structure for project information in general, and specifications and bills of quantities in particular. The principle is that work sections for bills and specifications are organised in a way that is much more closely related to the way in which work is executed and/or subcontracted on site[33].
❑ *Project Specifications: A Code of Procedure for Site Works*[34].
❑ *Production Drawings: A Code of Procedure for Site Works*[35].

The latter two documents give detailed recommendations for the production of drawings and specifications in a more organised and co-ordinated manner.

❑ *Co-ordinated Project Information for Building Works – a guide with examples*[36]. This document provides a comprehensive overview of CCPI recommendations and conventions.

There is little evidence that the recommendations of the CCPI have been widely adopted. A limited survey, at De Montfort University, Leicester, in 1992[37], suggested that only 55% of respondents were aware of the recommendations, and only just over a third had adopted common arrangement principles. The reasons given for this were various, but generally revolved around a reluctance to change existing practices.

Hand-over and commissioning

The roles and responsibilities of the various parties to the project continue into the occupation of the building, and it is pertinent to consider the information needs of the person taking delivery of the building. The Architects' Job Book[38], in stage K, includes an item requiring the architect to remind all relevant personnel to provide the necessary information for the preparation of a maintenance manual. It is a matter of some conjecture as to how often such a manual is actually produced.

Rooley[39] comments that:

> 'Commissioning and the attainment of the Practical Completion Certificate are therefore viewed from opposing standpoints by the designer and facilities manager. One has the expectancy of relief and escape from the project, and the other anticipation and excitement of a new beginning.
>
> Hand-over documents should include the design intent, an accurate set of as-fitted drawings, a set of commissioning figures and modifications carried out during commissioning and a set of maintenance instructions which are specifically relevant to the building and to the operation of each component in context.'

The CIOB refer to project information manuals[40]. Such a manual consists generally of two elements. One part will be concerned with provision of information to the building owners to enable them to use and manage the building properly, and the second part will consist of instructions for its management, with the broader facilities management scenario in mind.

Contents of the building manual

(1) Information

The most essential component is a set of as-built drawings. The *RIBA Job Book*[38] advises that these should be specially prepared. There are two major reasons for this. In the first instance, even the latest most up-to-date set of production drawings are unlikely to be as-built, as all instructions and variations will not necessarily be included in a revised drawing, and because of possible discrepancies between drawings and construction. Secondly, as-built drawings need to be produced for the very specific purpose of assisting with the proper running of the building, both for operational and technical purposes. Any attempt to 'fudge' the issue by the presentation of a pack of rather dubious information to the building owner will be counter-productive.

Possession of a relevant and accurate set of as-built drawings by the building manager is essential at a variety of different levels.

❑ At the strategic level all good maintenance management systems require accurate and up-to-date records of the estate. The as-built drawings, provided at hand-over, are only the first step in this process, as a continuous up-dating process is necessary to incorporate additions and alterations.

❑ These drawings need to be appropriately annotated for the specific purpose of managing the building. For example, information for fire safety and the needs of maintenance management require a particular format.

❑ At the operational level there will be the need for detailed information with respect to specific parts of the building system. An important example is the need to provide drawings identifying service routes, and the means of accessing them.

There is no standard pattern for the production of these drawings. The best advice that can be given is to stress that their preparation should be considered as a logical development of the project information system. The drawings should be properly co-ordinated with other information sources, because the drawings will need to be supplemented by other information, which may be in the form of schedules or tables.

The manual should contain all the necessary information about the construction of the building, and therefore comprise not only as-built drawings and schedules, but all the necessary information to ensure that proper management of the building, in the widest sense, can be effected. It is impossible to present an exhaustive list at this point, but each of the following aspects are worthy of consideration.

❑ A properly produced set of as-built drawings supplemented by relevant schedules, notes, and specifications of materials and components used in the construction of the building.

❑ A reference section, where all relevant sources of information are given. This may include manufacturers' and suppliers' data, including spare part numbers for plant and equipment, and guarantee or warranty information. British Standard references may also be relevant.

❑ A detailed breakdown of the parties involved in the design, construction and commissioning of the building. For example the addresses of consultants, sub-contractors, etc. should be included, along with points of contact.

❑ Historical data with respect to defects and their making good in the form of a log-book may provide important intelligence for the building manager.

If one adds to this collection of items other information required for managing the building, it very quickly becomes apparent that the information content of the manual is considerable, and represents in itself an information management problem of some complexity.

(2) Instructions for the use of the building
The need for occupiers to be aware of how to use a building properly has been highlighted in studies of energy usage. At the level of the simple dwelling, it has been found that sophisticated energy saving designs are to no avail, unless behavioural aspects are taken into account. This is clearly a significant factor when complex active service systems incorporating comprehensive control devices are used. Occupants must be adequately instructed as to their use.

There is a message here for designers, both in the assumptions they make when designing a building, and in the need to explain to clients how the building has been designed for use. For example, many modern dwellings incorporate conservatories as a sunspace. Whilst the theory behind them is perfectly sound, their benefits are negated when occupiers use them as an additional room all the year round, which entails heating them during the winter.

The building owner also needs to know the maintenance requirements of the building in terms of what to maintain, when to maintain, and how to maintain. However, it can be difficult to determine the level of detail that is required. This would be much easier to resolve if matters such as the organisation's attitude to maintenance policy and the nature of the building's usage have been accurately identified at the briefing stage.

For example, the building may become part of an established estate, with a clearly worked out policy in terms of planned maintenance, and the building manual would be written with this in mind. In such a case it might be entirely superfluous to produce a comprehensive maintenance guide in terms of planned maintenance recommendations. The manual can simply refer to an existing estate document, in which case only those maintenance features peculiar to the building under consideration need to be identified and accompanied by specific instructions.

On the other hand, it may be felt that the building requires its operations manual to give full and comprehensive recommendations for all its maintenance operations. However, to indulge in such an operation as part of the manual is probably a mistake as it will lead to the manual becoming far too cumbersome. There are exceptions to this, the most common of which relates to the mechanical services in the building. As well as making a heavy demand in terms of information content for the manual, it is this subject that will require the most attention in the instructions section of the manual, which has to be read in conjunction with the information section. Very careful structuring is essential to permit easy and accurate cross-referencing.

Preparation of the manual

Two problems immediately present themselves: firstly, as to how this information should be collected, and secondly, how it should be structured to facilitate its use by a diverse range of users.

Part of the answer to each of these questions depends on the way the project has been managed in its earlier stages. There are undoubted benefits to be derived if the collection of information for a building manual is seen as an on-going process throughout the project. Indeed, one could argue that the first set of data for inclusion, produced at the outset, is a proper set of user requirements.

The importance of the manual should be stressed, not only for its essential utility, but also for its tendency to act as a catalyst. At the brief development stage, the knowledge that such a manual is to be produced provides a great stimulus for proper consideration of how the building is to be managed, and therefore by implication the needs of the building and its users. At hand-over, the assembly of a package of information to accompany the building also encourages a critical analysis of what has been achieved. This can help feedback, but more importantly perhaps, if the preparation of the manual has been an integral part of the whole process, it should have led to a more considered approach to both design and construction, from the point of view of the eventual user.

Structure of the manual

There is no set formula for the structure of the manual. The two major components will, however, certainly require further sub-division, and there is a strong case for considering the information portion in two parts. This should consist of essential working documentation, such as the as-built drawings and schedules, which may then be further sub-divided into sections representing parts of the building or site.

More detailed reference material, such as manufacturers' data, may best be included in a specific reference section, which can be accessed from both of the two major sections. The case for separating out the reference section is a strong one, as this may be organised in the manner most appropriate for referencing from the other sections, rather than sub-divided into parts of the building.

The instructions may be sub-divided into either parts of the building or site, or into elements. A case may be made for either approach, and the decision in this respect will depend on the particular characteristics of the project. For example, in a housing project an elemental approach is most appropriate, whereas on a complex multi-functional development, sub-division may best be into sections of the project, or perhaps by building type.

To summarise then, a typical manual would consist of the following:

(1)   Instructions for use of manual

(2)   Information section
- ❑   General information
- ❑   As-built drawings and schedules
- ❑   Basic materials and components
- ❑   Parties to the project

(3)   Instructions for operation of the building
- ❑   General buildings management
- ❑   Maintaining the building fabric
- ❑   Maintaining the building services

(4)   Reference section

Presentation

The way in which this information is presented is open to choice. The use of loose leaf binders is convenient and facilitates easy updating. For a large building, several volumes will be necessary. However, considerable potential exists for storing the information by electronic means. This may range from the use of an integrated business package, containing spreadsheets and databases, with hard copy drawings, to the use of CAD systems, or even specialised facilities management software.

A number of large organisations are currently engaged in transferring all drawn information onto CAD systems, which as well as being efficient storage media, are also excellent data managers. The delivery of a building along with a set of such files from the design team clearly provides great potential. Linkage to databases, to form an integrated system, creates further opportunities to develop an automated building model, which may encompass full asset registers.

The help desk is another development where individual facilities are linked to a central information source, and is designed for use by non-professional, as well as professional staff. Future developments are likely to lead to the use of expert systems to help manage buildings, and these systems may also provide an excellent tool for providing feedback to design teams.

Maintenance and the building manual

A disciplined approach to the preparation of a building manual should generate the following benefits.

(1) It enables a better dialogue to exist between the designer and the maintainer, and also has the effect of promoting some feedback to the design team. If the maintainer has been involved at the brief development stage, as is advocated, then the preparation of the manual should follow quite naturally.

(2) It enables a property to be maintained more effectively, both in the organisational sense, and to the proper technical standard.

(3) It enables and encourages the building owner to plan the effective maintenance of the building, both in terms of planning maintenance programmes, and assisting in the formulation of budgets. It may, in addition, provide an important tool for the maintainer when they seek maintenance funding.

(4) By helping to ensure that the building is used properly, it contributes to the reduction of avoidable maintenance tasks.

(5) As a discipline for the design team, it fosters a more rigorous consideration of the effectiveness of the building in use, and may encourage a critical appraisal of how well intentions were defined, and then met. In this sense it provides some sort of testing ground for the assumptions made at the earliest stage of the project. If the original brief is developed in full expectation of a building manual being produced at hand-over, this will undoubtedly act as an important stimulus to give more complete consideration of performance, including maintenance. Similarly, during detailed design it may serve to encourage a rational approach.

The preparation of the manual does not, of itself, ensure a longer term involvement of the design team in its long term performance, but does act as an important focus at a critical phase in the process. The question of this longer term involvement by the designer has always been a tricky one, and in reality is probably not a feasible option.

It is all the more important, therefore, that the essential continuity is provided by some other person. This may be an in-house professional client, when one is available or, failing this, the maintainer or maintenance manager.

Maintenance feedback

Based on a database he has collected of building occupation problems over a 15 year period, Rooley[41] discovered that 60% of the buildings studied had apparent maintenance problems due to:

'... "fuzzy edge disease" where there is a lack of communication among the various parties in the design, procurement and maintenance process.'

This highlights the essential communication problem identified earlier, and the provision of sensible and reliable feedback information is one component that should be considered.

Harlow[42] comments that, without feedback from the industry as to how buildings have performed in the past, designers are unable to determine how their design decisions will work. This may take a variety of forms but, as pointed out by Clive Briffet[43], for it to be effective there has to be:

- ❑   An incentive to prepare it
- ❑   An expertise to analyse it
- ❑   A system to store it
- ❑   A readiness to use it

Maintenance activity generates vast quantities of data, and there are efficient means available for its storage and analysis. This of course may not be feedback specific information, although Holder[44] describes a system for recording user information for projects, that has been used on the Lloyds Building, and Williams[45] outlines the use of a building log for recording information.

It is important, also, to recognise that the information required not only relates to detailed performance behaviour but to the validity of building concepts. The most effective way of course is the involvement of building managers in the design process, preferably at briefing stage and, in some large estate owning organisations, this may be possible.

There is, however, a need for more published information, in the right format, and for this information to be taken note of.

'Feedback information is a continual problem and further efforts are required to ensure that problems found during maintenance works to have resulted from incorrect detailing, poor specification and choice of materials, defective construction techniques and workmanship, are not repeated, as these prove to be very costly. In this connection it would be necessary to record these defects and also to record the remedial measures which have been taken[46].'

References

(1)   Local Authorities Management Services and Computer Committee (1981) *Terotechnology and the Maintenance of Local Authority Buildings.* LAM-SAC, London.

(2)   British Standards Institute (1984) *BS 3811: 1984 Glossary of Maintenance Management Terms in Terotechnology.* HMSO, London.

(3)   O'Reilly, J.N. (1987) *Better Briefing Means Better Buildings.* HMSO, London.

(4)   Markus, T.A. (1977) *Building Performance – Building Performance Research Unit, Strathclyde University.* Applied Science Publishers, Barking.

(5)   National Economic Development Council (1978) *Construction for Recovery.* National Economic Development Office, London.

(6)   Powell, J., Cooper, I. & Lera, S. (1984) *Designing for Building Utilisation.* E. and F. Spon, London.

(7)   Stonehouse, R. (1983) Housing in the Humanist age, *Architects' Journal.* 9 February.

(8)   Catt, R. (1986) Bringing buildings into care, *Building Technology and Management.* 10 July.

(9)   Wallace, W.A. & Then, D.S.S. (1987) Maintenance considerations during the design process, In Spedding, A. (ed.) *Building Maintenance and Economics – Transactions of the Research and Development Conference on the Management and Economics of Maintenance of Built Assets.* E. and F. Spon, London.

(10) Lee, R. (1987) *Building Maintenance Management,* 2nd edn, Blackwell Science Ltd, Oxford.

(11) Building Research Establishment (1975) *Digest 176 – Failure Patterns and Implications.* HMSO, London.

(12) Powell, J., Cooper, I. & Lera, S. (1984) *Designing for Building Utilisation.* E. and F. Spon, London.

(13) New Builder (1993) Complex specifications force up UK building costs, *New Builder.* 16 April.

(14) Marsh, N.G. (1979) The effect of design on maintenance. In *Developments in Building Maintenance* (E.J. Gibson, ed.). Applied Science Publishers, London.

(15) Bowyer, J.T. (1985) Managing Building Maintenance. In Harlow, P. (ed.) *Building Maintenance – the Architect's Viewpoint.* CIOB, Ascot.

(16) Statutory Instruments (1995) *Health and Safety: The Construction (Design and Management) Regulations 1994.* HMSO, London.

(17) European Commission (1992) *Directive on Temporary or Mobile Work Sites – 92/57/EEC.* EEC.

(18) Health and Safety Executive (1995) *Managing Construction for Health and Safety: Construction (Design and Management) Regulations 1994 – Approved Code of Practice.* HMSO, London.

(19) Ashworth, A. (1989) Life Cycle Costing – a practical tool, *Cost Engineering*. March.

(20) Flanagan, R. & Norman, G. (1989) *Life Cycle Costing for Construction*. Surveyors Publications for the RICS, London.

(21) Palmer, A. (1989) Construction Design Cost Optimisation, *Chartered Quantity Surveyor*. February.

(22) Flanagan, R. et al. (1989) *Life Cycle Costing for Construction*. Surveyors Publications for the RICS, London.

(23) Ashworth, A. (1994) *Cost Studies of Buildings*. Longman, London.

(24) Lee, R. (1987) *Building Maintenance Management*, 2nd edn, Blackwell Science Ltd, Oxford.

(25) Mishan, E.J. (1982) *Cost Benefit Analysis – an Informal Introduction*. Allen and Unwin, London.

(26) Pilcher, R. (1973) *Appraisal and Control of Project Costs*. McGraw Hill, Maidenhead.

(27) Flanagan, R. (1989) *Life Cycle Costing – Theory and Practice*. Blackwell Science Ltd, Oxford.

(28) Zimmerman, L.W. & Hart, G.D. (1982) *Value Engineering – a Practical Guide for Owners, Designers and Contractors*. von Nostrand Rheinhold, New York.

(29) Co-ordinating Committee for Project Information (1987) *Co-ordinated Project Information for Building Works – a guide with examples*. CCPI, London.

(30) Building Research Establishment (1981) *Current Paper 7/81 – Quality Control on Construction Sites*. HMSO, London.

(31) National Economic Development Council (1992) *Achieving Quality on Building Sites*. NEDO, London.

(32) Institute of Advanced Architectural Studies (1982) *Design Decision Making in Architectural Practice – IAAS research paper*. IAAS.

(33) Co-ordinating Committee for Project Information (1987) *Common Arrangement of Work Sections*. CCPI, London.

(34) Co-ordinating Committee for Project Information (1987) *Project Specifications: A Code of Procedure for Site Works*. CCPI, London.

(35) Co-ordinating Committee for Project Information (1987) *Production Drawings: A Code of Procedure for Site Works*. CCPI, London.

(36) Committee for Co-ordinated Project Information (1987) *Co-ordinated Project Information for Building Works – a guide with examples*. Building Project Information Committee.

(37) Searle, M. (1992) Project Information for the Construction Industry – is Co-ordinated Project Information the Answer? Leicester Polytechnic, Undergraduate dissertation.

(38) Royal Institute of British Architects (1988) *Architects' Job Book. Volume 1: Job administration*, 5th edn. London.

(39) Rooley, R.H. (1992/3) Building Services: Maintenance Systems and Policies, *Structural Survey*. Winter.

(40) Chartered Institute of Building (1990) *Maintenance Management – a guide to good practice*. CIOB.

(41) Rooley, R.H. (1992/3) Building Services: Maintenance Systems and Policies, *Structural Survey*. Winter.

(42) Bowyer, J.T. (1985) Managing Building Maintenance. In Harlow, P. (ed.) *Building Maintenance – the Architect's Viewpoint*. CIOB, Ascot.

(43) Briffet, C. (1987) Maintenance Design and feedback – successes and failures. In Spedding, A. (ed.) *Building Maintenance and Economics – Transactions of the Research and Development Conference on the Management and Economics of Maintenance of Built Assets*. E. and F. Spon, London.

(44) Holder, W. (1989) Software based user information, *Building Services*. January.

(45) Williams, P. (1989) Keeping a record, *Chartered Quantity Surveyor*. April.

(46) Stavely, H.S. (1989) Maintenance management – the consultant's viewpoint, *Chartered Builder*. Nov/Dec.

Chapter 4

The Nature of Maintenance Work

Much maintenance work will be inevitable, as it is in the nature of materials to deteriorate over time with usage and exposure to the elements of climate. However, the rate at which the deterioration of materials and components takes place may, to some degree, be controlled by prudent decisions being made during the design stage of the procurement process.

Previous chapters have underlined the importance of having properly defined performance standards, against which a building's actual performance can be measured, and considerable development has taken place in the methods available for carrying out condition surveys of existing buildings. A well developed professional building surveying expertise now exists to interpret condition data, often by reference to an extensive body of knowledge on the performance of materials and components, in order to formulate and execute appropriate repair and maintenance strategies.

Despite the presence of this expertise, many property owners do not take professional advice about their buildings. For example, whilst the owner of a factory may investigate the efficiency and energy consumption of the manufacturing plant and machinery, he will rarely require a similar exercise to be carried out for the building fabric, as this is not perceived as contributing directly to productive output.

This means that, whilst it is quite feasible to create a condition model of a building, few building owners avail themselves of the advantages that such a model could bring in devising programmes of repair and maintenance. This short-sighted attitude results in inefficient buildings and potentially higher running costs in the longer term.

Routine maintenance

Routine repairs and remedial action

Routine maintenance, in its purest form, can be thought of as being work that has to be carried out at intervals in order to keep the building in an appropriate condition. This work may involve either the repair or replacement of an item, and is generally necessitated by natural deterioration or normal wear and tear, caused by a wide range of agencies. Common examples of routine maintenance are quinquennial (five-yearly) external decoration and the annual servicing of boilers.

The question to be addressed here is not if maintenance should be carried out, but rather 'what' and 'when'. For many routine operations there is sufficient data and/or experience to determine a reasonable cycle, and a carefully prepared maintenance manual will supply this information. Once this has been defined then budgets can be framed accordingly, and the work executed on a planned cyclical basis.

However, for problems other than those associated with abnormal or unreasonable deterioration, there are factors which mitigate against this ideal. For example, the correct advice may not be available at hand-over so that:

❑ All routine maintenance items have not been properly identified
❑ All routine maintenance items have been identified but the wrong advice has been given on the timing and nature of the work required

On the other hand, all the correct advice may be available but the building owner chooses not to take it. For example, he may perceive that increasing the length of maintenance cycles is a way of reducing overheads in the short term, and use this strategy, either to give a market edge, or increase profitability. In the public sector it may simply be part of a cost cutting exercise. The inevitable result is that items that should have been routine become non-routine through postponement to some indeterminate time in the future. Furthermore, there is a multiplier effect, in that failure to carry out routine maintenance at the right time is one of the factors that contributes to abnormal deterioration.

Another reason for not carrying out timely maintenance stems from a lack of knowledge rather than a failure to provide or take notice of information. A design team may, quite conscientiously, produce maintenance guidelines based on the information that is available. However, in general, there is a deficiency in data availability, both in terms of quality and quantity. Whilst it may be unrealistic to expect these information needs to be always met from within an organisation, it might reasonably be hoped that data generated nationally should be more helpful. This was recognised by the Audit Commission, whilst carrying out its 1986 English House Condition Survey[1], and they commissioned a report from NBA Construction Consultants, which was published ahead of the main report in June 1985[2]. This is probably the most comprehensive review and simple set of guidelines and bibliography on this subject yet produced. It summarises published knowledge and opinion, derived from 7000 titles and summaries. The authors of the review, however, comment in their introduction that:

'The details of maintenance cycles have to be set properly within the context of the maintenance objectives of individual organisations and the policies resulting from these, particularly the balance between planned and responsive maintenance.

The life expectancy data given here is at best indicative, in many cases representing averages from experience.

In any individual case, element lives will be dependent on many factors, such as design, quality of construction, climatic conditions or degree of use.'

Replacement of components

Initially, it is convenient to focus on replacements that might reasonably be expected, rather than those that may be termed avoidable. For example, there are items that are replaced on a routine basis as a preventive measure, i.e. it is accepted that the item has a finite life and is replaced before this is reached. This may be a short, medium or long term operation, although it will most commonly relate to items which have a limited life and are therefore replaced, rather than repaired or subjected to routine maintenance. However, even items with a medium or long term life may ultimately need to be replaced, having been repaired several times during their life, and/or subjected to routine maintenance.

It might be expected that routine replacement items are identified at the outset, together with recommendations for replacement cycles. However, as well as the problem of data availability, there are a number of other issues to resolve in determining whether or not to carry out a replacement. This applies also to the determination of repair cycles and both situations are appropriate for the application of optimisation techniques. These issues are explored more fully in Chapter 6.

Planned and unplanned routine maintenance

Under ideal conditions, at the outset of a building's life, it would be possible to have the maintenance of a building completely planned into a series of routine maintenance and replacement cycles. In practice, routine maintenance and replacement may not happen because of failure by the design team to communicate the necessary information to the building owner. Even where such information is provided, the owner may lack the will to use it or choose not to because of uncertainty.

Despite a professional approach by all parties, unexpected failures of components will occur, and deterioration rates of materials or components may differ from those predicted. Whilst in many instances failure could reasonably have been foreseen and avoided, there still remains a high degree of uncertainty in the prediction process.

Published data will normally present only average values of deterioration rates or replacement cycles. Because it is not possible to be perfectly predictive, continuous monitoring of the performance of a building is necessary in order to update knowledge, and improve the reliability of the data already in existence. Routine maintenance programmes must therefore make provision for a system of regular inspection and reporting.

Planned routine inspections

Inspections may be categorised as:

❑ Routine inspections that will, perhaps, focus on a predetermined selection of items
❑ Major comprehensive inspections, probably on a larger temporal cycle, that consider the whole building or sections of it, and which may take a variety of forms depending on the reasons for which they are carried out

Routine inspections are normally carried out in order to identify items in need of repair or replacement, many of which, being unpredictable, are not readily programmed into a routine maintenance cycle. The determination of appropriate inspection cycles are, of course, no easier to predict than maintenance cycles or replacement periods, and are just as vulnerable to a lack of commitment by the building owner.

Whilst many inspection items, particularly those relating to machinery and plant, may be clearly prescribed, fabric maintenance is rarely the subject of precisely defined inspection cycles. It is to be hoped that, increasingly, inspection checklists will be provided for all the elements of a building, and not just for mechanical plant, as they provide the main means of updating condition data.

By using carefully designed checklists, which clearly set out the criteria for classifying the condition of particular elements, it is feasible, after some initial training, to use non-technically qualified staff to collect condition data. Such checklists should incorporate a simple coding system to assist recording, help consistency, and indicate when any necessary work should be carried out. In addition to recording items for inclusion in a repair programme, the checklist should provide, through the coding system, a means by which the inspector can 'flag up' any items beyond his/her competence to deal with, for a qualified professional's inspection and advice.

There is often scope for readily incorporating some of the inspection work into day-to-day maintenance operations. For example, when decoration is being carried out, part of the task allotted could be to inspect and report on rainwater goods. If use is made of this procedure, then some care must be taken to ensure that the reporting system employed facilitates the correct flow of information. In other cases inspection work may be carried out by supervisors as part of their general supervisory tours.

Properly designed inspection pro-formas should state, using a coding system, who is competent to carry out a particular inspection item, and the person actually executing the inspection should identify themselves, so that reports can also be sorted on this basis. This is important, as there are obviously cases where a specialist inspector is essential.

A systematic approach should also make it possible for the inspector to plan a method and sequence of inspection. Carefully pre-planned pro-formas, similar to those described below for condition surveys, help to standardise these procedures, and also lend themselves to automated methods, which are increasingly becoming the norm for managing maintenance data.

If the data is to be downloaded into a database, then the structuring of the pro-forma has to be consistent with the structure of the database being used. All of the automated survey techniques outlined later may be used, and many of the comments made with respect to their relative merits will apply here.

Large estates and buildings also require a methodology for locating items within the building, and the building within the estate. This is normally achieved by means of some form of coding.

To facilitate retrieval, collation and analysis, the facilities and elements inspected should be capable of being grouped by all or any combination of the following:

❏ Location, either the building or parts of buildings
❏ Elements such as external walls or windows
❏ Result of the inspection, so that items for urgent repair may be identified easily and quickly

This clearly identifies the potential for the use of electronic databases.

Remedial maintenance

Technical failure

Failure to meet an acceptable standard may be as a result of the normal agencies of deterioration, but equally may be attributable to technical or managerial shortcomings, and the work necessary to restore the standard in such circumstances is classed as remedial rather than routine.

A number of factors may contribute to a technical failure:

❏ Inadequate brief collection and development
❏ Poor detailed design
❏ Buildability problems
❏ Poor workmanship
❏ Lack of consideration for the execution of maintenance in the design
❏ An inadequate delivery/hand-over package
❏ Abuse of the building, either in its use or its management

A substantial amount of time may be spent analysing maintenance and trying to apportion responsibility for a poor performance. Such studies help to provide essential feedback data, but will only be really productive if they are considered to be an integral part of the building design and the process of producing the building. If performance data is to be accurate, it is important to trace failures back to their source, as might reasonably be expected when maintenance work is identified through a comprehensive building survey.

Design and construction failures

These may be considered in two distinct categories. Firstly, failures which occur soon after completion, termed 'burn-in failures'. These are frequently accepted as an inevitable part of running-in the building, and are largely attributable. Secondly, there are items that fail over a longer period of time, and it is not always possible, or realistic, to properly identify their source. These may be termed 'useful life' or 'wear out' failures, although they cannot always be considered to be 'normal'.

Between these two will be on-going failures which can be classed as 'in-service'. In statistical terms, an analysis of failures produces a frequency distribution called a bath tub curve (figure 4.1), and this may be useful for planning purposes.

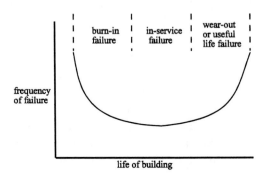

Figure 4.1 Bath tub frequency distribution curve for component failures.

However, overlying any consideration of construction failures is the issue of defining precisely what constitutes a failure. Useful life or wear out failures may often be subtly related to poor decisions made with respect to the choice of components or finishes. These failures are rarely recognised as such, and therefore create deficiencies in the accuracy of feedback data. A large number of failures are thus heavily dependent on perceptions. For example, if an item deteriorates at a more rapid rate than is reasonably expected, it may be classed as a failure by the informed analyst, even though it may not be perceived as such by the user of the building. Many of these failures present further problems, in that they may be latent and not manifest themselves for a number of years, by which time they may have had an adverse effect on some other component or element. This underlines the wisdom of an effective planned inspection policy to detect potential failures at an early stage.

Poor maintenance

It is possible to attribute a significant number of construction failures to poor maintenance practices, such as:

❑ Inadequate routine maintenance
❑ An ineffective replacement programme
❑ Lack of proper inspections on a planned basis
❑ Inadequate data to enable the preceding items to be properly carried out

Classification of maintenance

A classification of maintenance simply into routine or remedial, or planned and unplanned categories is clearly of rather limited value. The Audit Commission[3] considered a better division of maintenance to be as follows.

(1) Strategic repairs and maintenance
This represents work required for the long term preservation of an asset, and includes planned maintenance of the building fabric (decoration and routine replacement), maintenance of engineering services installations, and major repair items such as re-roofing. They are normally items which can be planned for because, to some extent, they can be foreseen and budgeted for.

(2) Tactical repairs and maintenance
These items relate to day-to-day work, of a minor nature, in response to immediate need. The Audit Commission point out that 'tactical maintenance' is not necessarily the same as responsive maintenance, as some immediate response items are clearly of a strategic nature, for example a flat roof failure.

BS 3811[4] definitions

The following definitions are all given in BS 3811 and, for practical purposes, it is clear that the maintenance work load will consist of a mix of all of these (figure 4.2).

(1) **Planned maintenance:** This is maintenance organised and carried out with forethought, control and the use of records to a predetermined plan.

(2) **Unplanned maintenance:** *Ad hoc* maintenance carried out to no predetermined plan.

(3) **Preventive maintenance:** Maintenance carried out at predetermined intervals, or corresponding to prescribed criteria, and intended to reduce the probability of failure, or the performance degradation of an item.

(4) **Corrective maintenance:** Maintenance carried out after a failure has occurred, and intended to restore an item to a state in which it can perform its required function.

(5) **Emergency maintenance:** Maintenance which it is necessary to put in hand immediately to avoid serious consequences.

(6) **Condition-based maintenance:** Preventive maintenance initiated as a result of knowledge of the condition of an item from routine or continuous monitoring.

(7) **Scheduled maintenance:** Preventive maintenance carried out to a pre-determined interval of time, number of operations, mileage, etc.

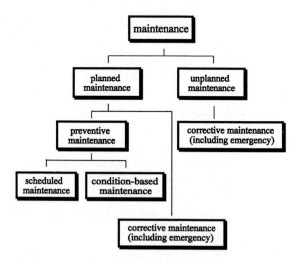

Figure 4.2(a) Types of maintenance.

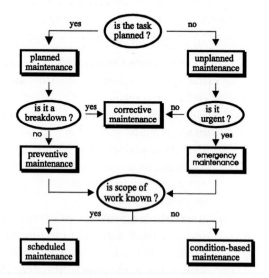

Figure 4.2(b) Decision based types of maintenance.

Estate records

It is not possible to manage a property portfolio effectively without an accurate picture of what the estate comprises, the characteristics of the properties and their condition. How comprehensive such information needs to be will be determined by management imperatives, and attitudes to maintenance. A prime requirement, however, is always to have an accurate global property record.

An estate record will consist, in its simplest form, of a list of all the properties, their geographical location and a description of their general characteristics. These characteristics should include occupancy information, technical information concerning the construction of the building, and, hopefully, a property history that is accurate and up-to-date. Typically the following might be included:

❑ The location of the building
❑ Landscaping and external works
❑ General construction information
❑ Age and broad condition description
❑ Services information
❑ Area and volume of the building
❑ Type of accommodation
❑ Type of usage, current occupancy, etc
❑ A brief history of the building including any alterations or additions that have been carried out and when
❑ Any proposals for future use or alterations
❑ Current value of the property

Such a property record will then form part of the estate or portfolio record, within which it may be given a reference number or code. This information can normally be obtained quite readily, stored in an accessible form, and, with good management, be easily kept up-to-date.

At some stage a detailed condition survey will be required for each property, necessitating a full, rather than a simple inspection.

Condition surveys

The Audit Commission[5] have been critical of the local authority practice of estimating maintenance expenditure by simply taking a notional percentage of the property value, and urge the use of proper condition surveys to derive more accurate estimates of maintenance expenditure. Condition surveys should, however, be commissioned for more than just budgeting purposes, as they have a wider application in the managing of building condition.

A major obstacle to carrying out the first comprehensive survey is the expense. On a national scale the UK building stock possesses very poor condition records and this represents a massive impediment to developing good maintenance management practices. Some progress has been made in recent years, particularly with respect to local authority buildings, where the prompting of the Audit Commission has had some effect. Within the private sector, however, there is still a startling reluctance amongst property managers to commit funding and commission detailed condition surveys of their buildings.

Even within the public sector, many of the condition surveys now being carried out are strictly limited in their scope. In many cases they are carried out for very specific purposes, usually related to financial management, rather than as part of a professional approach to managing building condition. This is very wasteful of scarce resources as, instead of one comprehensive condition survey being carried out, which could be kept up-to-date by periodic inspection, several *ad hoc* surveys are undertaken, the data from which is impossible to unify into a comprehensive picture.

Building surveys

A building survey is defined by the RICS as:

> '... making an inspection of buildings of all types, so far as the nature, design and structure of the building and conditions of occupancy and furnishings allow, and preparing a report expressing an opinion on the condition and standard of construction and recommending the extent of any necessary repairs and future maintenance requirements.'

The level of detail required of the survey will depend on the purposes for which it is carried out. The main types can be categorised under the three following headings, in generally increasing order of detail and hence cost:

❏ Valuation survey
❏ Pro-forma survey and report
❏ Structural survey and report

In planning and carrying out any survey, the three following interrelated issues need to be considered:

❏ Survey procedures and collection of data

❑ Presentation and structuring of data
❑ Data storage

A great deal of difference exists between carrying out a one-off survey and surveying a large portfolio of buildings. When preparing a property database for maintenance management purposes, data storage will be an important and potentially expensive factor. This has prompted increasing interest in cost-reducing survey techniques, which at the same time collect data in a form compatible with some overall management information system.

At the simple level, use of a pro-forma type survey is obviously attractive, as it enables a degree of standardisation. However, this approach requires very careful planning and consideration of the requirements at the outset, and is only realistic if there is the possibility of using the pro-forma a number of times.

Databases used for the storage and manipulation of large quantities of data are ideal for planning purposes, and provide an excellent tool for analysis which, if properly integrated within the management system, can be of great benefit in facilitating the transfer of information. If a database is to be used, important issues have to be resolved with respect to the structuring of the information. Building condition data for a group of houses lends itself rather easily to a database structure based simply on construction elements, all of which repeat themselves a number of times. In a more heterogeneous property portfolio, though, the question of the database structure is by no means a simple one.

The nature of the database structure, once decided on, will determine survey procedures and the preparation of pro-formas[6]. Although the principles for traditional manual surveys have been firmly established over the years, there are still many schools of thought with respect to the exact procedures to be followed, and the automation of any part of the process seriously calls into question the appropriateness of these traditional approaches. Developments in electronic data collection and storage require surveyors to re-appraise their methods at all levels.

Whatever approach is taken, the principal aim of a building condition survey is to provide an accurate picture of the property stock. In general the condition survey requires:

❑ An accurate description of each building element, which describes both the construction and the materials used
❑ An assessment of the state of repair for each element, together with a statement of obvious defects
❑ An indication of the expected physical life of each element

- ❏ A schedule of repair work required, and an indication of the priority attributable to each item
- ❏ A cost estimate of the work required, which may not be provided at the time of the actual survey, but as a follow-up exercise

The amount of detail and depth of coverage required for each element under consideration is extremely variable, and in practice will be determined by the precise requirements and purposes of the survey.

In terms of data collection four approaches can be identified:

- ❏ Manual
- ❏ Optical mark reader
- ❏ Bar-code reader
- ❏ Hand-held computer

Manual methods

Manual surveys are frequently conducted using a pro-forma, which acts as a guide to the information to be collected. This may range from a checklist or series of prompts, to a rigidly structured survey form. The former may give more flexibility to the surveyor in terms of his survey technique, but this is often offset by the need to restructure the information to suit data storage needs. Whilst this approach lends itself very well to the collection of factual material, it cannot be a substitute for the judgemental aspects of the surveyor's work, such as assessing condition and useful life of a component.

Many of the surveys currently being undertaken by local authorities for maintenance management purposes are very objective in that they are highly structured rather than judgemental. This is a result, not only of financial constraints, but also, of a limited set of survey objectives (figure 4.3).

The surveyor's knowledge and experience will always be vital, and a systematised approach can never give a full picture, and will certainly be weak diagnostically. Notwithstanding this, the major problem encountered with the manual approach is the difficulty of maintaining consistency between surveys and the surveyors carrying them out.

In considering the production of planned maintenance programmes, consistency is necessary and this may be achieved by using more sophisticated methods.

Optical mark reader

This method uses a pre-printed form, designed specifically for the task in

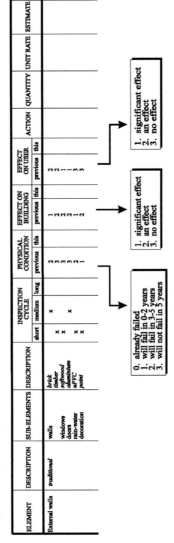

BIENNIAL SCHOOL SURVEY - CONDITION SURVEY PRO-FORMA

SCHOOL - Greenbridge Junior School

SURVEYOR - N. Parker

DATE - 15 March 1995

ELEMENT	DESCRIPTION	SUB-ELEMENTS	DESCRIPTION	INSPECTION CYCLE			PHYSICAL CONDITION		EFFECT ON BUILDING		EFFECT ON USER		ACTION	QUANTITY	UNIT RATE	ESTIMATE
				short	medium	long	previous	this	previous	this	previous	this				
External walls	traditional	walls	brick		x		2		1		2					
		windows	timber	x			3		2		2					
		doors	softwood	x			3		2		1					
		rain-water	aluminium	x	x		3		2		1					
		decoration	uPVC	x			2		1		3					
			paint				1		2		3					

0. already failed
1. will fail in 0-2 years
2. will fail in 3-5 years
3. will not fail in 5 years

1. significant effect
2. an effect
3. no effect

1. significant effect
2. an effect
3. no effect

1. significant effect
2. an effect
3. no effect

Figure 4.3 Example of condition survey pro-forma.

hand, which consists of a series of multiple choice questions. The reply to each question is marked in a space provided next to the appropriate option, using a graphite pencil. When the form is completed, the survey sheet is fed into an optical mark reader (OMR) for processing.

The survey sheets are printed in a colour that the computer based reader is programmed to ignore. It therefore only reads the graphite pencil marks, interpreting them according to their position on the sheet. In most applications the information is stored in a database, which may also contain a built-in schedule of rates, permitting rapid assessment of cost implications (figure 4.4).

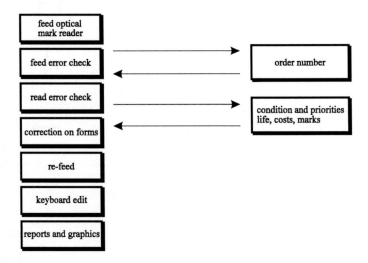

Figure 4.4 Procedure for feeding survey forms to OMR.

The consistency of the information gathered is improved drastically, in this method, but this has more to do with the use of a structured survey sheet than the technique *per se*. The possibility of using an in-built schedule of rates is extremely valuable, and may be applicable in any of the methods, provided the information is input in the correct format, and can easily be kept up to date.

Difficulties can arise when so called 'rogue' items, that are not included in the schedule, are observed. These will need to be priced manually, which can be very disruptive when the whole work environment is electronically based. Programmes can be written that provide built-in checks for consistency and gross errors, but the systems are limited by the extent to which contingencies are provided for, both on the form and in the software. The two have to be compatible, and hence there may be a tendency to lose accuracy in the interests of expediency.

Bar-code reader

This method uses pre-printed multiple choice forms, as described above, but, instead of a mark, each question is allocated a bar-code. As the survey is carried out the surveyor uses a light pen against the bar-code for the appropriate answer. The light pen is connected to a hand-held portable computer which stores the data in a database format as the survey proceeds. On completion of the survey the portable computer downloads the data to the main computer in the office.

The merits of this technique are very similar to the previous, and the major advantage is in the reduction of labour at the data transfer stage. On the other hand the equipment used is more cumbersome.

Hand-held computer

A hand-held computer can be programmed, prior to the survey, with a series of questions that can be multiple choice, or devised to act as prompts. The surveyor answers the questions as the computer asks them, so the order in which they are asked is crucial. Built-in checks can be programmed which, being operative during the survey, provide a major advantage. These checks may include a requirement that no question can be missed, and range checks that prohibit the entry of absurd answers. After the survey the results can be downloaded from the hand-held computer as before (figure 4.5).

The major advantage of using hand-held computers is the flexibility they provide over pre-printed pro-formas, by using standard software which can be customised to suit the needs of each survey.

A contemporary system (figure 4.6) makes use of a hand-held computer, together with a bar-code reader. Much attention has now been given to the ways in which programs can be written and used to structure questions and data, and to provide for efficient input and output.

Reports

The information collected from a building survey can be presented in a number of ways. It can be in the form of a written report, in a spreadsheet format, or visually in the form of charts and graphs (figures 4.7 and 4.8). The spreadsheet is appropriate to individual bespoke surveys, but for maintenance management purposes, charts, schedules and graphs are more likely to be appropriate.

What will have been obtained in all cases is a picture of the condition of the buildings comprising the estate. The major cost is in executing the

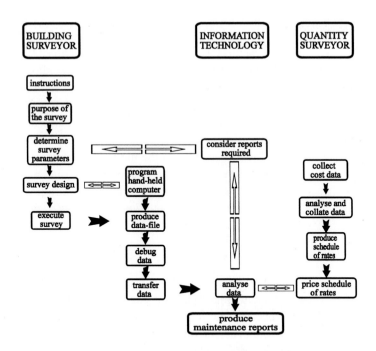

Figure 4.5 Conducting a survey using a hand-held computer.

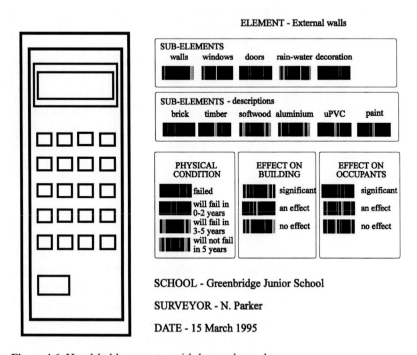

Figure 4.6 Hand-held computer with bar code reader.

BIENNIAL SCHOOL SURVEY : Summary of findings on elemental basis

SCHOOL : Greenbridge Junior School

SURVEYOR : N. Parker

DATE - 15 March 1995

ELEMENT	SUB-ELEMENTS	GENERAL CONDITION	ACTION	EST. EXPEND. (£)					FUTURE ACTION
				year 1	year 2	year 3	year 4	year 5	
Substructure	strip foundations concrete slab blue brick DPC	all good (visual inspection only)							
External walls	brickwork timber cladding	good fair	decorate and monitor	400	50	50	500		possible replace
Windows and external doors	softwood aluminium	fair good	decorate and monitor	500	50	50	500		possible replace
Roofs	traditional - pitched flat - felt rain-water	good fair fair	renew clean out	75	75	350 75	100	100	
Internal fabric	blockwork stud partitions	good fair	repair	200					
Internal finishes	walls floors ceilings paintwork	good fair good fair	replace carpet decorate		800 750				possible repairs
Services	gas water electricity internal drainage fittings	good(visual inspection) ditto ditto ditto fair	detailed inspection ditto ditto ditto replace			50 100 50 100 250	250		possible future work on all services following inspection
External works	paths steps walls external drainage roads & car parks landscaping	good fair fair fair fair good	repair repair clean resurface	300 100	400 100	100 1000	200	200	

Figure 4.7 Summary of condition survey recommendations.

first survey, and in many instances the picture obtained may be limited to that required by immediate demands, and thus often only provides sufficient information for budgeting purposes.

Applications

School maintenance budgets

School maintenance represents a major challenge for local authorities. A typical approach taken by the estates department is to survey all schools under their control to produce a condition weighting factor for each building, which is then used to produce a weighted school population. Such a weighting factor is based on many criteria, such as the age of the school and its type of construction – information that can be obtained without recourse to a condition survey.

In one typical LEA condition surveys were carried out with two purposes in mind:

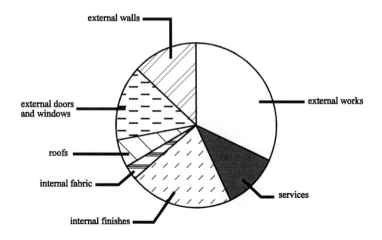

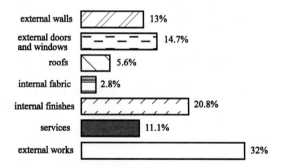

Figure 4.8 Use of pie charts and bar diagrams.

(1) To give an overall assessment of the school's condition in order to produce weighted school population figures for:
  ❑ Devolved budget allocation
  ❑ Organisational planning

(2) To identify major repair items for forward planning purposes.

The National Health Service

A very structured approach has been taken within parts of the NHS, where the move to trust status has necessitated a thorough evaluation of capital stock, including buildings. Linked with this evaluation has been an imperative to assess the usage of every building thoroughly. The approach used is described below, and is one which considers building condition as part of an overall asset management exercise. The detailed methodology varies from authority to authority, and the one outlined is based around that described by Sahai[7].

Value for money is the over-riding issue, and the approach is consistent with Audit Commission guidelines. The exercise is carried out as a property appraisal, with the following factors considered:

- ❏ The function of the building
- ❏ Location
- ❏ Utilisation
- ❏ Suitability for purpose
- ❏ Condition
- ❏ Energy efficiency
- ❏ Fire/safety risk

In other, similar, approaches the last of these items might be broadened and specified as statutory requirements, and perhaps include insurance requirements.

In the property appraisal, each of the items is evaluated and described by the allocation of a code. For example, under utilisation, the following designations are provided:

(1) Empty or grossly under used
(2) Under used
(3) Adequate
(4) Overcrowded

Under energy efficiency there could be:

A. Ideal
B. Adequate
C. Does not meet building regulations
D. Major change required
X. A suffix when replacement is the only option

In considering building condition, the evaluation is rather low key, and only requires the allocation of one of the following designations:

A. As new and can be expected to perform adequately to its full normal life
B. The element is sound, operationally safe and exhibits only minor deterioration
C. The element is operational, but major repair or replacement will be required within three years for a building element, and one year for an engineering element

D.   Serious risk of breakdown

X.   The suffix to category C or D for when replacement is the only option

The appraisal of physical condition would normally be carried out for each of 19 physical elements. Typically, these might include:

❑ Structure
❑ External fabric
❑ Internal fabric
❑ Roof
❑ Electrical installation
❑ Mechanical installations
❑ External works

For every section of the property appraisal, a weighting is given to each category so that the scores under each can be combined to give an overall score for each of the buildings surveyed. In terms of physical condition, it is also possible to combine the assessment of all 19 elements into a single condition score for the building. For this purpose, each of the 19 elements has to be given a sub-weighting totalling one. For example, within this the structure may be given a weighting of 0.25.

This is essentially what may be termed the asset management approach, and has as much to do with accounting and financial management as maintenance management, although it does have the merit of highlighting the importance of building condition within a wider scenario. In providing an overall score for the building, physical condition is given a weighting of 0.24 compared to 0.59 for functional suitability. Of interest here is the weighting given to safety and statutory standards of only 0.10.

The problem in this system is that the categories used for the appraisal are not mutually exclusive, and it should be borne in mind that much of the work carried out on the physical condition of the building will be necessitated by statutory requirements and this will influence functional suitability.

Local authorities

Large housing authorities, and county council property departments, have adopted a systematised approach to building condition for a number of years but, more recently, property records have become more operationally based, in the sense that they provide some sort of basis for formulating a maintenance plan.

The Audit Commission[8] recommend that all local authorities carry out a systematic review of maintenance requirements on a periodic basis. They quote as an example Cumbria County Council, who carry out five-yearly surveys. Their work is also described by Bargh[9], and in a report produced by the DES[10].

The methodology requires buildings to be broken down into sets, the criteria for allocation to a set being:

❏ Period of construction
❏ Materials used
❏ Type of construction
❏ Style

The objective is to produce sets of buildings that have similar basic characteristics, so that their condition can be analysed and compared in the most meaningful way. The advantage that then flows is that the survey requirements of each set can be carefully formulated, and this may lead to economies in terms of execution of the surveys, enhanced appropriateness of pro-formas, and some standardisation in the form of the data obtained.

In the early reviews, surveys of all properties looked at 11 basic elements, e.g. roofs, floors, external walls, services, etc. These elements were further subdivided into individual features such as structure, insulation, drainage, decoration, etc. and each feature given a score of from 1 to 6, representing dangerous to excellent.

The purpose was to identify maintenance needs on a more scientific basis, but the approach is now regularly adopted, the major differences being in the classification system adopted. For example, Colston has described a method used by the SSHA[11], which was developed in response to a perceived continuing decline in government support for housing maintenance, within an organisation where the majority of the housing stock was more than 40 years old. This necessitated the derivation of sensible priorities and a property appraisal was carried out with this in mind.

To do this, each housing development was assessed in terms of the work required over the next 10 years for 15 major elements of the internal and external fabric. This was complemented by a follow-up planned inspection programme, not only to identify work needing to be carried out over the period, but also to assess priorities. A priority coding system, in terms of the degree of urgency, was used.

For example, public safety was assessed on whether it would be a danger within five years, in which case it was allocated a priority [1] code, whereas, if it was likely to be a danger after five years it was given a priority code of [3]. Similarly, fabric repair items were classified according to whether the work was a replacement, due to the component reaching the end of its useful life, or whether it was to extend its useful life. The former would be given a higher priority. There are many other variations on this theme, all of which tend to be geared to short term objectives within an organisation. It is evident that for maintenance purposes, condition surveys are rather systematised affairs and therefore produce structured data.

Despite the fact that the surveys are often carried out for very specific motives, the data produced will almost inevitably have other uses, particularly as a highly structured format lends itself to storage in some sort of database format on a computer. The usefulness of the data stored is then very dependent on the efficiency with which it can be managed.

The evaluation of buildings by qualified surveyors, to determine their condition and hence assess the potential maintenance work, is operated under the presumption that professional judgements are made. In making these judgements, the surveyor should have knowledge of, and take into account, any specific requirements laid down by statute.

Statutory requirements

The legislative framework for building standards is principally concerned with matters affecting health and safety[12], and is set out in Acts of Parliament, which are supplemented by detailed regulations that are published from time-to-time. The Acts range from those that relate to buildings in general, to those that cover the condition of specific types of building.

The legislative provisions can be divided into those which control the design and physical requirements of new buildings, alterations and extensions, such as Town and Country Planning Acts and the Building Act, and those which relate to the occupancy of a building. The latter mainly come under the provisions of the Health and Safety at Work etc. Act 1974 and its regulations, which have progressively replaced a range of other Acts over recent years.

Planning Acts

The major principle of these is to control the use to which land is put, and covers not only new buildings, but changes to existing buildings, either in their physical characteristics or their use. These can then, in some circumstances, have relevance to the execution of major repair work, where planning requirements may control the nature and appearance of any work carried out.

Of greater importance under this heading, however, are the powers given to the Secretary of State to compile lists of all buildings which are of historical or architectural merit. Listed building consent must be obtained from the local planning authority before any work is carried out on such a building.

Thus far, planning can be seen as a controlling influence on what is carried out and how it is done, rather than a generator of work. However, it is worth noting that a local authority has the power to prevent an owner allowing a listed building to deteriorate, by serving him with a Repairs Notice specifying the work required to preserve the building.

The Secretary of State also has powers to make grants for the maintenance and repair of buildings of 'outstanding' architectural or historic interest, and for the preservation or enhancement of character or appearance of conservation areas.

Building Act 1984

This Act consolidated much previous primary legislation relating to the physical requirements of buildings, including the 1984 Building Control Act which introduced fundamental changes in the method of building control. There are several parts to this Act.

(1) Part I is concerned with Building Regulations. Under s. 5.1 the Secretary of State is empowered to make regulations with respect to the design and construction of buildings and the provision of services, fittings and equipment, for the purposes of:

- ❑ Securing the health, safety, welfare and convenience of occupants and other people affected by the building
- ❑ Furthering the conservation of power and fuel
- ❑ Preventing waste, undue consumption, misuse and contamination of water

S. 5.2 is of particular importance in that it provides that Building Regulations may impose continuing requirements on the owners and occupiers of buildings, including those which were not, at their time of erection, subject to regulation. These may apply to what are termed designated functions, such as keeping fire escapes clear or, in respect of some services, fittings and equipment, such as the periodic inspection and maintenance of lifts.

(2) Part II relates to the arrangements for supervision of work other than by local authorities.

(3) Part III gives local authorities the power to require the owners or occupiers of buildings to provide certain facilities, or remedy certain deficiencies, and it may give notices requiring work in connection with the following typical items:

- Drainage outside the building
- Sanitary provisions in a building used as a workplace
- Quality of sanitary provision
- Means of escape in the case of fire in consultation with the fire authority
- Raising of chimneys
- Alteration or closing of cellars and rooms below sub-surface water levels

It may also serve notice on an owner or occupier that it intends to carry out certain actions itself, for example:

- where a building is in such a state, or is used to carry such loads, as to be dangerous
- the repair, restoration or demolition of a building which is, by its ruinous or dilapidated condition, seriously detrimental to the amenities of an area
- paving and drainage of yards and passages

In the following cases the local authority may apply to a magistrates court for an order to:

- Temporarily close or restrict the use of a building, pending the provision of satisfactory means of ingress and egress
- Execute remedial work to a dangerous building or restrict its use, where danger arises from overloading

In some cases, persons carrying out certain types of work are required to give the authority a prescribed period of notice before starting the work. This occurs, for example, in the case of altering the course of an underground drain.

Building regulations

These apply to building work in England and Wales, and to certain changes of use of an existing building. Equivalent regulations apply specifically to Scotland and Northern Ireland. Building work is defined as:

- ❏ The erection or extension of a building
- ❏ The material alteration of a building
- ❏ The provision, extension or material alteration of a controlled service or fitting
- ❏ Work required on a material change of use

The regulations say nothing specifically about the point at which repair becomes subject to control. Normally this would not be controlled, but if the point is reached where a building has been seriously damaged or dilapidated, the authority may reasonably apply the regulations to any remedial work.

In general the Building Regulations will be relevant only to the nature of a repair, or the way in which it is carried out, rather than a generator of maintenance *per se*.

Housing Acts

These are of prime importance when maintaining housing stock. Indeed, in many situations, where resources are scarce, they may provide the minimum standard that a housing authority provides, and become the major maintenance planning yardstick.

There are a range of Housing Acts that lay down standards relating to residential premises and these relate to:

- ❏ The condition of the fabric – repair, stability, dampness, natural lighting and natural ventilation
- ❏ The equipment and services – sanitary fittings, hot and cold water supply, drainage, cooking facilities, artificial lighting and heating installations
- ❏ The internal layout – space for activities and circulation and privacy in houses in multiple occupancy
- ❏ The quality of the surrounding environment – air pollution, noise level, open space and traffic conditions

The Housing Act 1957, re-enacted in 1985, laid down criteria for determining whether or not a house is unfit for human habitation. A house may be adjudged unfit as a result of shortcomings in respect of one, or the combined effect of two or more of the following.

(1) Repair
The state of repair should not be a threat to the health of, or seriously

inconvenience the occupiers. The internal decoration is not taken into account here. The problem still remains a judgemental one in this case. In the first instance, it is not easy to demonstrate a direct link between ill-health and disrepair and, in the second, it may not be clear as to what constitutes serious inconvenience.

(2) Stability

There should be no indication of movement which may constitute a threat to the occupants. Observation over a period of time may be necessary to establish this criterion, and it ignores the fact that not all structural failures are associated with early signs of distress.

(3) Freedom from damp

Dampness should not be so extensive as to be a threat to health. Dampness is a consequence of disrepair in many cases, although design faults may also be contributory. One way or another, some diagnosis is required, and the problem remains of establishing a direct link to ill-health.

(4) Natural lighting

There should be sufficient light for normal activities, under good weather conditions, without the use of artificial light. No absolute standard is laid down, but case law suggests the use of a sky factor of 0.2%. There are serious problems of enforcement, and the issue is complicated by matters such as privacy and rights to light.

(5) Ventilation

There should be adequate ventilation to the outside air in all habitable rooms and working kitchens. The guidance contained in Approved Document F to the Building Regulations provides a sound benchmark with respect to current best practice.

(6) Water supply

There should be an adequate and wholesome water supply. There are a plethora of guidelines with respect to this, and the issue of water quality is becoming increasingly controversial.

(7) Drainage and sanitation

There should be a readily accessible WC in a properly lighted and ventilated compartment.

(8) Facilities for preparing and cooking food

There should be a sink with an impervious working surface, a piped water supply, cooking appliance and waste water disposal.

(9) Internal arrangement

This requires that the internal layout of the house should not constitute a hazard or cause serious inconvenience.

Defective Premises Act 1972

The purpose of this Act is to impose duties in connection with the provision of dwellings, and to attribute the liability for injury or damage caused to persons through defects in the state of premises. The Act has potentially far-reaching implications for landlords and building owners.

In the first instance, the Act imposes a strict duty to build properly. This duty is owed to the person for whom the dwelling is provided, and to every person who subsequently acquires an interest in it. A person who does work to any premises is under a duty at common law to take reasonable care for the safety of others who might reasonably be expected to be affected by defects arising from their work.

Where premises are let under a tenancy which obliges the landlord to maintain and repair the building, the landlord owes to all persons who might reasonably be expected to be affected by a defect, a duty to see that they are reasonably safe from personal injury or from damage to their property caused by the relevant defect. This duty is owed by the landlord if he knows, or if he ought in all circumstances to have known, of the relevant defect. There are clear implications here in terms of repair and maintenance for the landlord and, a direct encouragement to him to carry out regular inspections of his properties.

Factories Act 1961; Offices, Shops and Railway Premises Act 1963

The detailed legislation in these Acts continues in operation until subsumed under the more unified Health and Safety at Work etc. Act. The expression 'factory' is defined as any premises in which persons are employed in manual labour, in any process for, or incidental to, a range of prescribed activities. The duty of observing the provisions of the Act falls primarily on the occupier, although the owner will incur a measure of liability in some cases. Offices, shops and railway premises are also defined in the Act, and again, the main duty for observance falls on the occupier.

The Health and Safety Executive is now responsible for enforcement of both Acts, except where there are express provisions placing this duty on local authorities. The enforcing authorities may appoint inspectors, who have wide powers to enter premises, and to make such investigations as may be deemed necessary.

The general requirements of the Acts are very similar, and are concerned with the safety and welfare of employees working in the building, rather than its contents. The main provisions affecting maintenance relate to:

- Cleanliness and, in the Factory Act, painting
- Overcrowding, by reference to minimum space requirements
- Temperature
- Ventilation
- Lighting (this also requires that sources of natural and artificial light are maintained at the appropriate standard and turned on when required)
- Sanitary conveniences, their provision, cleaning and maintenance
- Washing facilities
- Drinking water
- Accommodation for clothing
- Seating arrangements
- Eating facilities
- Floors, passages and stairs must be properly constructed and maintained, and their safety ensured

The terms 'properly maintained' and 'reasonably practicable' appear many times throughout these Acts. Properly maintained means, for the purposes of the Acts, maintained in an efficient state, in efficient working order and in good repair. The adverb 'properly' would suggest a normally accepted standard of good workmanship, with particular emphasis on safety.

The need to ensure that all these provisions are properly complied with places a premium on regular inspection. Much of this may be of a routine nature, and the setting up of a proper management information system is highly desirable. The use of CAD systems, with linked databases to form a management information system, is becoming increasingly attractive.

Health and Safety at Work etc. Act 1974

This Act provides a comprehensive and integrated approach, which governs the safety, health and welfare of people at work, and the members of the public who may be affected by such work.

Part I of the Act defines the people upon whom general duties fall and these include:

- Employers
- Employees and self-employed people

- ❑ A person in control of non-domestic premises used by persons not in their employ
- ❑ Designers, manufacturers, importers, installers, or
- ❑ Suppliers for their products

This part of the Act provides for the creation of a Health and Safety Commission, and its Executive, which is the main body responsible for enforcing the statutory requirements. Enforcement powers are also given to local authorities and other bodies for certain aspects of the Act. The Executive will appoint inspectors to administer the provisions of existing legislation and any new legislation that may be enacted. Amongst their powers, they have the authority to issue improvement and prohibition notices.

Part II re-enacts the Employment Advisory Service Act, within which the Advisory Service advise on matters relating to the safeguarding and improvement of the health of employed persons. The detailed regulations under the Act are referred to as the Health and Safety Executive's 'six-pack':

- ❑ Management and Health and Safety at Work Regulations 1992
- ❑ Provisions and Use of Work Equipment Regulations 1992
- ❑ Health and Safety (Display Screen Equipment) Regulations 1992
- ❑ Workplace (Health, Safety and Welfare) Regulations 1992
- ❑ Personal Protection Equipment at Work Regulations 1992
- ❑ Manual Handling Operatives Regulations 1992

To these, and of particular relevance to property professionals, can be added the Construction Design and Management Regulations 1994, discussed in Chapter 3.

The maintenance manager may have much of the responsibility for protecting the interests of the occupier, in providing safe and healthy working conditions, as the Act places a responsibility on the person in control of the premises to ensure that not only are the premises themselves safe, but also the plant and machinery therein. Thus, the maintenance manager, depending on his defined role, could incur personal liability for non-compliance.

The maintenance manager must, therefore, consider himself as being not only in a position where he needs to respond to requests for work to remedy a defective situation, but also one in which he must act in a proactive way, so that compliance with the legislation becomes part of a programme of planned inspection and maintenance.

The safety duties imposed can be summarised as follows, and the implications for proper building maintenance should be carefully noted:

- ☐ Make and keep all work places safe
- ☐ Ensure the safety of machines and materials
- ☐ Plan and use safe working systems
- ☐ Plan, train and direct employees
- ☐ Receive and consider employee views
- ☐ Ensure the safety of subcontractors and visitors
- ☐ Protect the well-being of co-tenants, neighbours and the public at large
- ☐ Test and supply in safe conditions any goods for use by employees or others
- ☐ Effectively safeguard all employees wherever they may work
- ☐ Set out in writing the arrangements and organisation to do this

Enforcement measures include criminal prosecution, prohibition or improvement notices, industrial tribunal hearings, and civil claims based on negligence or breach of statutory duty.

It is clear that there are a host of legal matters with which the maintenance manager needs to be acquainted, some of them operational in terms of their implications. A great many matters that fall under the responsibility of those managing and maintaining buildings, require action on a day-to-day basis, and must therefore be considered as part of a planned inspection system and thus generate maintenance work in the form of preventive measures.

In many types of building the need to comply with the law will be the major generator of maintenance, and may consume the major proportion of a maintenance budget. There is evidence to suggest that within the NHS this is already the case, with the result that the major expenditure is on plant and equipment, where potential breach of the law is perceived as having the most serious consequences. Fabric condition problems, from the legal point of view, may be rather longer term in nature, and often more difficult to substantiate.

Fire safety

Fire safety in buildings is controlled, either by reference to the Building Regulations, where new building work is involved, or the Fire Precautions Act 1971, when a building is already occupied.

The Building Regulations, administered by the building control authority, control all new building work, i.e. the erection or extension of a building, material alterations or material changes of use, and set out requirements for structural fire precautions and means of escape in case of fire.

The Fire Precautions Act 1971, administered by the fire authority, is concerned with occupied buildings, used for a purpose designated under s. 1 of the Act. Buildings used for a designated purpose require a fire certificate, unless they are exempted (exemption generally relates to small premises where there are few people). The designated uses are:

- ❑ Hotels or boarding houses
- ❑ Factories
- ❑ Offices
- ❑ Shops
- ❑ Railway premises

The primary duty for observing the Act lies with the occupier, although the fire authority may impose responsibility on others. Following a designating order, application must be made to the fire authority for a certificate. They will then inspect the premises and issue a certificate specifying its conditions, which will normally include:

- ❑ Use of the premises
- ❑ Means of escape and its maintenance
- ❑ Fire fighting measures
- ❑ Fire warning methods

Additionally, the certificate may stipulate certain requirements for maintaining all these provisions. Inspectors may enter buildings, and are responsible for reviewing certificates from time-to-time.

Obligations under tenancy agreements

The apportionment of responsibility, between landlord and tenant, to maintain a building subject to a lease, can be provided for by covenants in the lease agreement. In theory, this should be quite a straightforward matter, but in practice it can become quite a complex affair. A body of case law on dilapidations has developed over the years and this continues to grow.

The word dilapidation, in general use, expresses some notion of disrepair. When used by a lawyer or surveyor, it normally refers to a property that is not only in a state of disrepair, but where, as a consequence, a legal liability is incurred. This liability is normally incurred by virtue of a lease which has placed an obligation on either the tenant or the landlord to maintain and repair certain parts of the property. Liability may also be incurred, however, under the law of tort or by statute.

Waste

'Waste may be defined as an act or omission which causes a lasting alteration to the land or premises affected to the prejudice of the owner or landlord. Waste is a tort and independent as a source of liability from any contractual relationship expressed or implied, between parties, so that an obligation not to commit waste is not excluded by an express or implied covenant in a lease dealing with the same subject matter.'

West's Law of Dilapidations[13]

There are a number of types of waste:

- ❑ Voluntary waste – the deliberate carrying out of some act which tends to destroy or injure land, premises or landlord's fixtures
- ❑ Permissive waste – allowing buildings to fall into disrepair by neglect, and failure to repair them
- ❑ Any unauthorised change in the nature or character of premises is waste; however, if changes result in the betterment of them then this is termed ameliorating waste

Liability for waste is founded in tort, and the measure of damages is the amount of damage to the landlord's reversion, i.e. the depreciation in sale value of his interest, which may not be the same as the cost of reinstatement. Actions in waste are uncommon, but may be the only redress against persons in occupancy of premises without a tenancy, or other persons in occupation of land without express or implied obligations to repair.

Statute

Parliament has modified the common law rule of no implied liability on landlords to carry out repairs or see that premises are fit or suitable for occupation by the tenant, under the Landlord and Tenant Act 1985. This applies in the case of short leases of dwelling-houses, and imposes certain repairing liabilities on landlords which may not generally be contracted out of.

Landlord and tenant obligations under leases

The obligations to repair and maintain are normally laid down in repairing covenants. In a general covenant to repair, the tenant will undertake one or all of the following obligations: to put, keep or leave the premises in a defined state of repair.

A specific covenant to put in repair by the tenant would be unusual, since it is implied by a covenant to keep in repair, that the tenant will first put the premises into repair, if they are in disrepair. Where dilapidated premises are to be let, this type of covenant could be used to get the building put into repair early in the life of the lease. In such a case the covenant will usually specify a particular standard of repair, and time scales will be given for the necessary work.

With a covenant to keep in repair, it may be implied that the tenant will have received the premises in good repair at the commencement of the lease. The condition of the building at this time may be regarded as the benchmark of the standards required, and should, therefore, be the subject of a schedule of condition, agreed by both parties at the outset.

If the only repairing covenant is to leave the premises in repair, then no action can be brought against the tenant, during the lease, to carry out repairs. The landlord must wait until the end of the lease. This underlines the importance of establishing a schedule of condition at the creation of the lease.

There are a number of problematic issues, all the subject of a large body of case law:

❏ The meaning of repair
❏ Standards of repair
❏ Fair wear and tear

This is not the place to comment in detail as a thorough treatment of these issues is provided by Hollis[14].

Most leases contain a clause requiring the tenant to be responsible for carrying out works that may be required to comply with current statutes, and there are a number of possible remedies for breach of repair covenants. The applicability of each of the following depends on the nature of the lease and individual circumstances:

❏ Forfeiture of the lease, in which case the tenant may also be required to make monetary compensation
❏ Injunction to prevent a party to a contract from carrying out some act which he has been covenanted not to do
❏ A court decree compelling a party to carry out what he has been covenanted to do
❏ The landlord executes the repairs, which is only possible if there is an express provision in the lease for him to enter and execute repairs
❏ Where a tenant has covenanted to repair and failed to do so, then damages may be recovered, and this amount will be determined by whether or not the lease is still operative or has terminated

Condition surveys of rented properties are obviously, therefore, of particular importance, both at the beginning and end of a lease, and there may be a specific requirement depending on circumstances. Schedules of dilapidations may often be prepared, frequently by both parties, with the intention of defining repair obligations, rather than for use as a management tool.

References

(1) Department of the Environment (1988) *English House Condition Survey 1986*. HMSO, London.
(2) NBA Construction Consultants (1985) *Maintenance Cycles and Life Expectancies of Building Components: A guide to data and sources*. HMSO, London.
(3) Audit Commission for Local Authorities in England and Wales (1988) *Local Authority Property – A Management Handbook*. HMSO, London.
(4) British Standards Institute (1984) *BS 3811: 1984 Glossary of Maintenance Management Terms in Terotechnology*. HMSO, London.
(5) Audit Commission for Local Authorities in England and Wales (1988) *Local Authority Property – A Management Handbook*. HMSO, London.
(6) Building Maintenance Information Ltd/Pitt, T.J. (1990) *Expert Systems and Condition Surveys – Special Report 187*. RICS.
(7) Sahai, V. (1987) Management of built assets – value for money, In Spedding, A. (ed.) *Building Maintenance and Economics – Transactions of the Research and Development Conference on the Management and Economics of Maintenance of Built Assets*. E. and F. Spon, London.
(8) Audit Commission for Local Authorities in England and Wales (1988) *Local Authority Property – A Management Handbook*. HMSO, London.
(9) Bargh, J. (1987) Increasing maintenance needs on limited budgets. A technique to assist the Building Manager in decision making, In Spedding, A. (ed.) *Building Maintenance and Economics – Transactions of the Research and Development Conference on the Management and Economics of Maintenance of Built Assets*. E. and F. Spon, London.
(10) Department of Education and Science (1985) *Maintenance and Renewal in Educational Buildings – Design Note 40*. DES.
(11) Colston, A.E. (1987) Maintenance of public housing assets, In Spedding, A. (ed.) *Building Maintenance and Economics – Transactions of the Research and Development Conference on the Management and Economics of Maintenance of Built Assets*. E. and F. Spon, London.
(12) Ellis, P. & Tong, D. (1985) Health and safety in the office, *Facilities*. July.
(13) West, W.A. (1979) *The Law of Dilapidations*. Estates Gazette, London.
(14) Hollis, M. (1988) *Surveying for Dilapidations*. Estates Gazette, London.

Chapter 5
Information Management

Introduction to information systems

Maintenance operations require, and in turn generate, large amounts of data. The collection and processing of this data into management information is, therefore, a key issue in managing building maintenance. Whether manual or automated means are employed, there exists an information system in one form or another. In this sense information systems should not be confused with IT, as in many organisations they will still largely be manual.

Information management involves the design of a system in which data is collected and processed into management information. This makes several important distinctions in terminology, and emphasises that information is a function of data, a process, and the needs of the user.

A system can be represented simply, as shown in figure 5.1, in terms of input and output. Figure 5.2 represents a simple view of an information system related to condition surveys, and illustrates a method of describing the elements of a system. There are a number of issues that stem from this simple model.

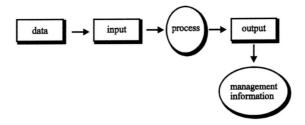

Figure 5.1 Information system in terms of input and output.

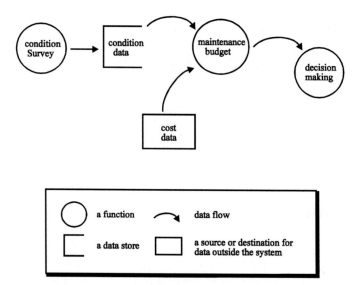

Figure 5.2 Elements of an information system.

(1) Raw data is subjected to processes that convert it into management information. In some instances this process may generate new data.
(2) Data may be collected and/or generated within the system, or may exist as an external entity.
(3) Information systems exist to serve user needs, which therefore need to be defined. The system must then be designed to produce information appropriate to these needs.

The information system can be viewed as having three essential components, relating to the organisation, data operations, and technology. In business terms this can be represented as shown in figure 5.3.

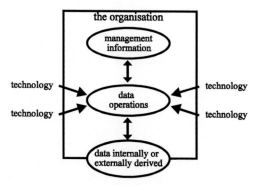

Figure 5.3 Components of the information system.

Information systems and the organisation

Organisations' structures grow into existence in a wide variety of ways, to satisfy a number of objectives. Information management will be one of these but not until recently can it be said to be a major determinant. The growth of IT, amongst other factors, has raised information management to a higher position on the business agenda, so that a systems approach to the design or development of organisation structures is now more likely.

This raises an important philosophical issue as to what should be the motive force for redesigning the information system and introducing new technology. It is possible to argue that information management is the major imperative, and that the organisation structure should be subjugated to its needs. Many would argue, though, that this runs the serious risk of neglecting human needs.

There is no perfect answer to this, as consideration has to be given to the characteristics of the organisation in general, and its information management needs in particular. The evolution of an information system strategy should be treated as a component of the overall strategic thinking of the organisation, rather than as an independent entity.

The development and implementation of such a strategy is a complex and delicate operation (figure 5.4), and the complexities of the process have inevitably increased with technological potential.

During the first generation of new technology, refinements to information management were driven by data processing requirements, and

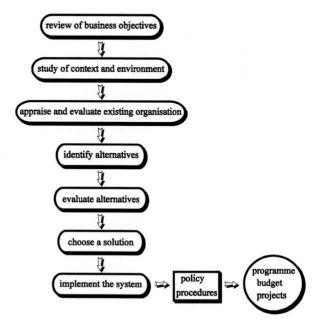

Figure 5.4 Development of an information system strategy.

developments that replaced previous manually executed functions, such as finance and accounts. The emphasis here was on efficiency, with management information somewhat incidental. In this instance the process can be classed as a bottom-up approach.

The development of more recent technology, and the potential for an integrated approach necessitates a much more strategic approach, and a top-down process (figure 5.5)[1]. For the future of maintenance management, this could have far-reaching implications, as the increasing tendency for maintenance management to be subsumed into facilities management leads to it being considered in a far broader strategic framework.

PROCESS	DRIVING FORCE	EMPHASIS
bottom-up	data processing	efficiency with information secondary
top-down	senior management	information needs determined by organisational objectives
systems evolution	data users	user orientated based on data manipulation
strategic development	top management	corporate strategy formulation with data/information integration

Figure 5.5 Implementation processes for information systems.

Information systems and technology

The development of information systems thinking is inextricably linked with the development of new technology and is perceived by professionals in terms of generations, which are paralleled within organisations by the introduction of ever more sophisticated data management applications[2]. The stages, or generations, that have been identified are outlined below.

(1) Manual records and filing systems.
(2) Early computerisation, which was used to replace existing manual systems, within a very contained organisational boundary, typically the accounts department, with technology based on main frame systems, and with specialised data processing departments.
(3) As applications increased there became a tendency for applications to cross boundaries leading, in some cases, to duplication and proliferation in an unstructured manner.

(4) The development from this is a result of the search for an integrated approach, and a more formal and effective information system.

(5) Computer-based systems increasingly began to take over other key aspects, such as the monitoring and control of resources. This was accelerated by a move towards user friendly hardware and software, and simpler data processing.

(6) Database concepts were developed that permitted data storage in a common filing system, with applications access on a controlled company-wide basis.

(7) There was a natural evolution to networked systems and mini-systems.

(8) There is now an ever-increasing demand for more efficient information flow, and a move towards the automated office.

These stages, of course, overlap considerably and represent a quite simplified view. The value derived at each stage is perceived, in most organisations, in terms of efficiency through stages 1 to 4. Stages 5 and 6 were, essentially, seen as contributing to an improvement in performance in strategic terms. Future developments will be driven by both efficiency and effectiveness, but inevitably greater convergence and integration of systems are likely.

Data operations

This represents the essential information management aspect, and requires consideration under a number of headings.

(1) Data sources

Data can be generated internally or externally, and may be formal or informal. Formal data sources can largely be designed into the information system in general, whether internally or externally generated. Formal data is also more likely to exist in a defined data store or data source, whereas informal data may be less well organised. Much of the effort in new technology is devoted towards the more efficient organisation of informal data.

Data sources can also be classified according to the point at which the data is generated, in which case it is sometimes categorised into primary, secondary or even tertiary data. This is of some importance in maintenance management.

(2) Data collection

In its basic form, data relates to a number of events and needs transmitting to a data store. In primitive systems this would be in handwritten form and transferred to a file. In the interests of efficiency a number of pieces of paper, or pieces of data, would be dealt with together leading to the term batch processing, from which derives the term batch input. Input to a data store normally requires some checking process to confirm its validity, and today this procedure is likely to be electronically based, via a data-preparation phase on a computer.

Increasingly, attempts are being made to by-pass the data preparation stage by the use of optical mark reading, optical character recognition and magnetic character recognition. Currently some research effort is being devoted to the codifying of maintenance data, to facilitate more effective input.

(3) Data output

Modern technology makes possible a variety of ways of organising and processing data and presenting it as management information. Central to this is the need to define management information requirements. It is not too easy to generalise here, but for decision making purposes figure 5.6 represents a hierarchical concept that is typical of many organisations[3].

Figure 5.6 Hierarchical decision making in the information system context.

Maintenance information needs

If maintenance is considered properly, in the context of the organisation, information needs should be defined at strategic, management and operations levels.

At the strategic level, the importance of developing an effective policy for the management of buildings requires the flow of appropriate management information in a corporate information system, especially with respect to building performance. It is a matter of some debate as to whether the maintenance of buildings often attracts this level of thinking.

It is at the operational level that most attention has been given to maintenance information systems and, in the majority of cases, integration with the organisational system and management decision making is confined to financial matters. The most common, and often the only link, is into the accounting system, and perhaps some convergence in the area of administration and office automation.

Most advanced maintenance management systems tend to be independent entities, so that as an information system, their development is somewhat stunted.

Accepting the generally perceived imperfections, it is still worthwhile examining maintenance information requirements in principle, throughout the whole life cycle of a building. Lee[4] adopts this approach by dividing the cycle into the design system and the user/occupier system. There is some merit, however, in extending the design phase to include construction, given the essential interdependence of the two. Similarly, the user/occupier system can be extended to embrace associated activity under the heading of the Building Management System.

The design/construction system

This phase should be considered to last from inception to commissioning and hand-over. It will possess its own internal information system (figure 5.7), through which data is collected and developed to produce what may

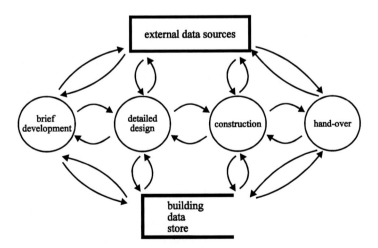

Figure 5.7 Information system during building procurement.

be termed a building data store. If this is carried out properly, which represents a major information management task, it should culminate in the presentation of a package of building management information at hand-over in the form of a building manual.

However, this information system is not independent, and must link into the overall organisational information system, and access external data sources (figure 5.8). The information needs can then be considered at three levels.

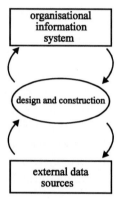

Figure 5.8 Design/construction and information systems.

(1)   User/client sourced material that permits:
  ❏ The development of a proper brief, that is an accurate response to user requirements, and which will include an explicit statement with respect to the stance to be taken on building maintenance
  ❏ A proper decision making process to take place, concerning the overall procurement strategy
  ❏ The physical execution of the design and construction process

(2)   There are additional information needs, over and above that provided by the user/client, to permit the development of detailed aspects of the brief, design and construction. Out of these very wide requirements there will be maintenance specific information such as:
  ❏ Maintenance cost information and component/material performance data to allow the evaluation of design alternatives
  ❏ Technical performance data/feedback to inform detailed design development
  ❏ Data for incorporation into the building manual

(3)   Information required specifically during the design/construction phase.

These needs can be classified in two ways. In the first place, data will be required from within the organisation, through proper feedback, which places a premium on a proper internal information system. Figure 5.9 illustrates a simple feedback loop with information flowing from a building data store into the design process. However, informal information channels may be just as important here. The link marked by a broken line could, for example, represent the involvement of a maintenance manager in the design process.

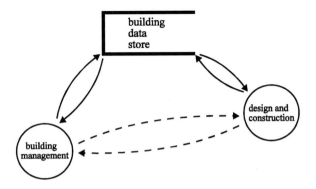

Figure 5.9 Maintenance feedback and design/construction.

Second, will be the requirement for data or information that can only be met by external sources. Although many of these are now well established, there is a major problem here of proliferation and a lack of a co-ordinated system for channelling it in the most effective way.

Taken in their entirety, these needs may be general in nature. For instance, there may be externally derived data with respect to the performance of a component. On the other hand it may be specific to the organisation, perhaps relating to an operational characteristic, or to the particular building being produced, such as a specific user requirement or activity. Specific information or data may be derived from internal and external sources, although the latter is only likely to exist with the presence of a large, well organised user group.

The building management system

The essential link into the building management system, from the design and construction phase, is the building manual, and it is to be hoped that this will more commonly become an integral part of the organisational information system.

In terms of managing the built asset, it is necessary to take the perspectives of both the building user and the building manager. Their imperatives, and hence their information needs, will be different. The former is concerned with optimising the use of the building and the latter is charged with the execution of specific activities that will satisfy that requirement.

In optimising the use of the building the user needs to know:

❑ How to use the building in the manner for which it was designed, assuming of course that this is consistent with the user's needs
❑ What action to take if it is felt that the building fails to meet these needs

The first of these should be addressed by a properly structured building manual, to which the user should have access. In practice there may be:

❑ No proper manual in existence
❑ A poorly produced manual
❑ A fundamental ignorance on the part of the user coupled, perhaps, with a reluctance to use what is available

This highlights the general need to increase awareness in the building user of performance issues, and specifically underlines the need to produce a building manual for both professional and non-professional users.

The action to be taken in the event of a perceived failure requires a reporting mechanism and, on the face of it, this is a simple interface with the building management function. This may not always be as simple as it sounds. At the level of the simple maintenance request, the flow of information is relatively straightforward. However, underlying these requests may be a more deep-seated dissatisfaction with the overall performance of the building (figure 5.10).

The interface between the user and the building manager can be seen as a conduit through which flow:

❑ Simple maintenance requests
❑ Data providing essential building performance intelligence, which needs to be added to the building data file and is an important component in the feedback loop for design teams

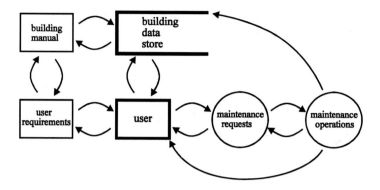

Figure 5.10 User requests and the building information system.

The latter of these is subject to a number of problems:

❑ The absence of a proactive approach by building users
❑ Failure to channel and direct data correctly
❑ Shortcomings in the analysis of the data into proper management information

There is a strong tendency for the building manager to be a passive recipient of data rather than an active searcher for it, other than to satisfy his immediate needs. The perfect feedback system on building performance cannot be effective without data, and it is unreasonable to rely totally on the user to provide it.

The manager of the building requires data or information to answer four fundamental questions (figure 5.11):

❑ What needs to be done?
❑ Can it be done?
❑ How is it to be done?
❑ When is it to be done?

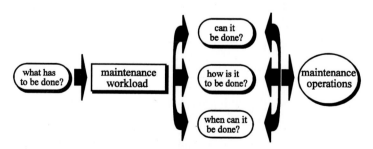

Figure 5.11 Generating maintenance tasks.

In determining what has to be done the building manager may identify the following categories:

- Requests from the user
- Needs identified during the manager's own investigations, e.g. inspections and surveys
- Regular items required by a variety of agencies, ideally taken from a building manual

There are therefore three sets of data flow, all of which constitute what may be termed the maintenance workload (figure 5.12).

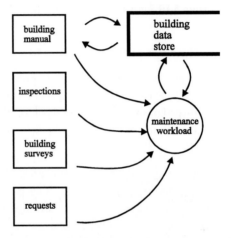

Figure 5.12 Generating the maintenance workload.

In response to this the building manager must decide what can be done, which may be limited by technical or financial considerations. These two are linked, but the latter, in particular, identifies the need for him to have up-to-date financial data. This requires him to be provided with a budget, and for there to be a flow of financial data permitting him to make decisions (figure 5.13). The essential requirements are to know:

- How much an item will cost, derived, hopefully, from a cost database section of the building data store
- The annual budget
- An up-to-date knowledge of what has been spent and committed
- A predicted financial out-turn

These factors, and the question of when, are inextricably linked, and are an essential part of the maintenance planning system. Access lim-

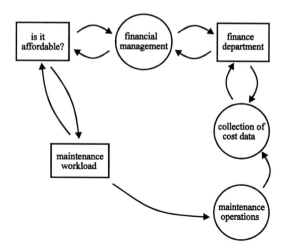

Figure 5.13 Maintenance and the financial information system.

itations, imposed by the use of the building, frequently constrain the timing of maintenance work. The information with respect to this may be obtained from the building manual, but can also be provided by the source of the workload item. For example, planned inspections, in identifying the need for work, can also specify access limitations.

In determining how to carry out an item of work, there are both technical and operational matters to consider. The first of these requires access to technical data, which may be available internally or externally. The collection of data internally to aid technical decision making is an important function of the information system, and it should be possible to refer to the building data store for this. There will also need to be sufficient information available so that a repair methodology can be devised, which includes overcoming physical access difficulties. This requires accurate up-to-date knowledge of buildings in the form of graphical information, as part of the building data store.

For operational purposes, the manager has to make a number of decisions ranging from the strategic, such as deciding whether to use direct or contract labour, down to the particular, such as deciding which contractor or maintenance operative to use. This is clearly an important part of planning, and the system set up must provide for a steady stream of information to assist the decision making process (figure 5.14). Additionally, the information system must provide for the proper flow of information to operative level and, following completion of the task, back into the system (figure 5.15).

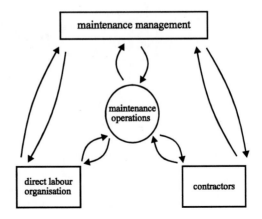

Figure 5.14 Maintenance operations information system.

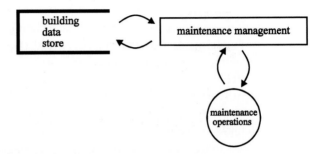

Figure 5.15 Operational feedback to building data store.

Maintenance information management

Building data store

The building data store referred to is, in effect, the information systems terminology for the building performance model. Figure 5.16 summarises the data flows to and from it, over the building's life. On a large estate it would be subsumed into a comprehensive model of the whole property portfolio. This store, or model, which is dynamic in nature, exists not only as the recipient of data generated but, also, as the essential internal source of data and is part of the system for generating management information. The close relationship of the building data store with the building manual is obvious, but they are separate entities. The manual should be seen as management information derived from the data store.

The means by which the data store is managed is a crucial issue. In

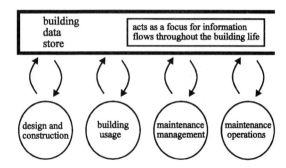

Figure 5.16 Inputs to building data store.

some situations it may quite justifiably be a purely manual operation but, beyond the most simple of estates, there is a strong case, and increasing pressure, for the introduction of new technology and the computerisation of the whole process. In the first instance the data handling power of electronic data management represents a massive technical advantage and, secondly, permits comprehensive analysis of data to produce the most appropriate management information. In addition, the process of introducing new technology imposes a disciplined approach to information management, and compels a critical appraisal of existing practices (figure 5.4).

Computer databases

Until recently, computerised maintenance management systems involved the processing of data using files, of which three types can be distinguished:

❏ A master file containing permanent information as a source of reference data
❏ Transaction or work files relating to physical operations, and which can be used to update the master file
❏ Report files, constructed by the extraction of information from the other files

The development of ever more powerful databases permits the production of fully integrated management systems. Three distinct types of database can be identified.

(1) Hierarchical databases link data in a tree-like structure. These are rather inflexible in use, but are extremely efficient for processing

large amounts of data. Their use is therefore most appropriate when there is a clear established information system in terms of input, output and data flows.

(2) Network databases are an extension of hierarchical ones but permit more cross-linking to improve flexibility.

(3) Relational databases are extremely flexible and are most appropriate for the varied requirements of managing maintenance information. In this type of database, information is organised into a series of files, but with data repeated in different files to create links between them. The very nature of maintenance information requires data to be duplicated in different groups, and a relational database permits this type of data organisation.

There are a large number of database processing applications in building maintenance. These range from off-the-peg systems, to execute specific tasks, which might be termed closed systems, to loosely structured database packages, which can be customised to suit individual requirements. The choice of a system is a complex issue, involving many factors, and specialist advice may be necessary.

The main advantages of such computerised systems are the rapid availability of information, and the ability to manipulate and organise data into a format that is appropriate to the maintenance team. A reduction in clerical work[5] is also likely, which allows the transfer of scarce resources to more productive effort.

The discipline imposed by the adoption of a computerised system encourages a systematic approach to the treatment of maintenance data. If the system is designed correctly it also provides a large measure of management continuity. There are a number of small maintenance organisations where the only information system is inside one person's head and this is clearly an untenable situation in the long run.

It is also clear[6] that, as with all computer applications, management commitment is essential if the system is to work. This must include, not only maintenance managers, but senior management in the organisation as a whole.

Integrated systems

More recent developments in computerised systems for managing maintenance have been towards fully integrated management systems, which provide direct on-line entry of job information, such as works orders, invoicing and payment details. This information then automatically updates property record files and contractor information. The general principles of such a system are shown in figure 5.17 and outlined below.

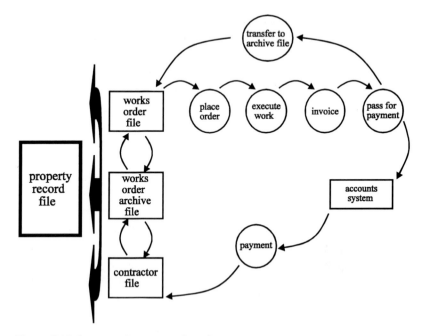

Figure 5.17 Integrated computer based system.

(1) Property record file
This file contains basic information such as:

❏ Tenant's name
❏ Address
❏ Property code

There are then several additional sets of information which may be included in the basic file, or in a sub-file accessed from the main property record file. There is some merit in the use of sub-files as it speeds up processing time. This information will include any or all of the following information:

❏ Construction details
❏ Age of property
❏ Details of services
❏ Cyclical maintenance or inspection requirements
❏ Maintenance history

(2) Contractor file

This file will keep the key information relating to currently approved contractors. There is a great deal of information that can be included and, again, it may be useful to sub-divide this file into permanent and variable information. The permanent contractor information will need to include most of the following information:

❑ Name
❑ Address and telephone number
❑ A code number
❑ Type of contractor
❑ Hourly rates including out of normal hours
❑ Tax certificate details
❑ Insurance details

It may also be useful to include some information with respect to the size of job that is appropriate, or the area within which the contractor operates. A typical sub-file file might be a record of the jobs undertaken by this contractor, permitting various analyses to be undertaken.

The example shown in figure 5.17 makes use of a works order archive file, and the data stored there may be retrieved in a number of ways, including on the basis of an individual contractor. This gives rather more flexibility, in that a code is used for the type of contractor. For example, all plumbing contract information can be extracted from this one file, allowing comparison of like with like. There are obviously a variety of ways of approaching this problem, and the one chosen will be that most appropriate to the organisation in question.

There are a number of programs that can be written to retrieve information from all these files for analysis purposes. Most standard database packages provide the facility to do this. Many of the so-called specialist maintenance packages are only relatively simply tailored databases using this facility.

A problem, however, with some off-the-shelf packages is that programming basically locks access to files, other than through the routines provided with the software. Thus, flexibility is limited, and there is a massive reliance on software specification at the outset.

(3) Works orders

These may be raised in a variety of ways. The system described has a program that will extract cyclical data from the property file and automatically raise an order. It is common for systems to raise orders via the computer and, in general, the following will need to be provided:

- Property address or code
- Contractor or contractor code
- Date
- Origin of order
- Estimated cost
- Start date and completion date
- Degree of urgency
- Details of work required

(4) Live orders file

When an order is processed, the details automatically get passed to a live orders file. There are a number of programs that can then access live order data to provide data for management control purposes.

(5) Invoicing

When invoice details are received from the contractor, the information is again entered into the system. The important information to be included is:

- Order number cross reference
- Amount of estimated and actual cost
- Date

In theory, the actual cost, date and order cross reference should be sufficient, but processing can be simplified by duplicating information. Once the invoice has been received and passed, the appropriate order is removed from the live orders file and the information contained in the order is transferred to update the property and contractor file. The order also then passes into an order archive file. This is normally a separate storage medium, such as a disk or magnetic tape, so as not to clutter up the live system with data.

There are a number of programs for accessing archive information. The degree of sophistication possible here depends on the adoption of a rigorous operational coding system. It must always be borne in mind what the requirements of the information management system are, as the need to use a complex coding system may neither be necessary nor realistic, given the circumstances of the organisation. A simple example of how file cross referencing may work is illustrated in figure 5.18.

The system described above can only be classed as an integrated information system, in that it does not fully facilitate full maintenance planning and control. Current developments are leading to the development of fully integrated management systems, and at the most

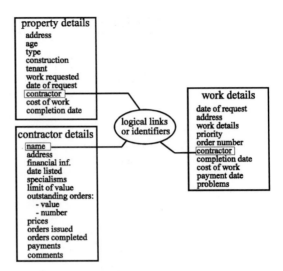

Figure 5.18 Cross-linking of data files.

sophisticated level to the incorporation of CAD systems, which can be viewed as graphic databases.

Fully integrated management systems

A fully integrated management information system, marketed as FrontLine™, has been installed at De Montfort University, in order to manage a rapidly growing building stock. The vast majority of the estate comprises of older buildings, for which records are by no means comprehensive. The system provides for full integration with a CAD system, but this step has yet to be implemented. The system can also be interfaced with a CAD driven facilities management system, the principles of which are described below. The incorporation of CAD into the maintenance/FM operation is necessitating a major exercise to carry out measured surveys of all buildings for entry into the CAD environment.

The basic components of FrontLine™ are shown in figure 5.19. The structure of the buildings, plant and equipment file, also referred to as the asset register, or index of items, is shown in figure 5.20, with its linkages to sub-files. Figure 5.21 illustrates the layout of the main page of the asset register or index. It should be noted that the index interfaces with planned maintenance files, which can in turn access a file of standard maintenance tasks that can be customised by the user.

As can be seen from figure 5.19, there are components for planned preventive maintenance (PPM), work planning and control, and stock control. It is worth noting that the asset register links with the PPM

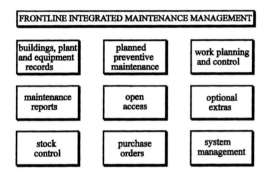

Figure 5.19 Elements of the FrontLine™ system.

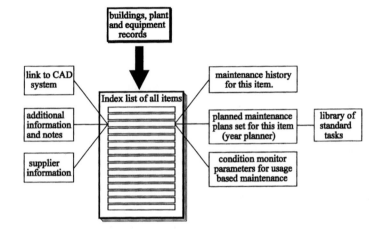

Figure 5.20 Structure of buildings, plant and equipment records in FrontLine™.

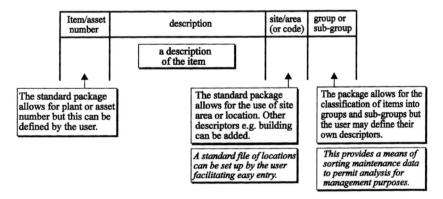

Figure 5.21 Equipment/asset register structure.

module via standard maintenance operations that are time related, i.e. through cycles, or by monitoring usage, e.g. metering of plant items. The system works on the relational database principle, and there is a separate module with a comprehensive range of management reports. These are shown in figure 5.22.

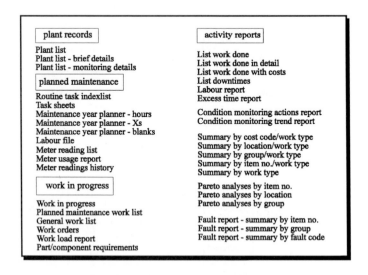

Note that Pareto analyses list headings in descending order of magnitude of cost, type, downtime or number of incidents with a cut-off specified.

Figure 5.22 FrontLine™ management reports.

CAD systems

Building maintenance operations require individual items of work to be carried out at prescribed locations, which are normally indicated on drawings. These may be in hard copy format, or stored on microfilm or microfiche. Whether they can be considered to be up-to-date is uncertain. In many cases they may only be the original as-built drawings, provided when the building was commissioned, and in others may be so old as to require complete redraughting, following a full measured survey. If the original drawings, or the new ones, are CAD based then their regular updating is greatly facilitated.

Alongside this graphical information will be a large body of text data, either in a building manual, on a computer database, or in an integrated system such as that described above. Consistency between graphic and text data is often very questionable, and there has more recently been a move towards a more integrated approach, particularly in the facilities management area[7]. By attaching database information about the building, or facility, to the graphic entities in the drawing, the database can be accessed via the on-screen drawing.

AutoCad™ is a particularly widely used CAD system, and many peripheral software packages have been designed to integrate with it. Its use in construction has been enhanced by the development of the Architect, Engineer, Constructor extension (AEC). The system acts as an electronic drawing board, and can be used in two or three dimensional mode. AEC allows the direct introduction of building components such as walls, roofs, windows and doors, selected from a library of standard components. In addition, many component manufacturers now produce product information on a disk, for incorporation into AutoCad™.

For a CAD system to be used effectively, the information contained in its files must be organised in a systematic manner. This is to ensure the efficient transfer of information between different members of the design, construction and management team. This, therefore, requires the adoption of a standard method of data organisation.

Part 5 of BS 1192 gives guidance on the structuring of computer graphic information, with the aim of simplifying its transfer. It identifies two methods of data organisation in CAD systems:

❑ The use of sub-models
❑ The use of a layering system

The use of the former requires the adoption of a hierarchical approach, which is less flexible than layering systems, that are now almost universally employed.

Sets of graphical items that make up a component, or part of one, are termed entities. When these are entered into a model, they are assigned to a layer. This can be visualised as a transparent sheet of paper. The full CAD model thus exists as a series of layers, or transparencies. Each layer can be activated or suppressed by the operator, on the screen, or in a plot.

Information can thus be grouped logically on an appropriate layer, and a high level of selectivity is available. As well as reducing clutter on drawings, this speeds up processing time and provides a very powerful information management tool. It is quite possible to name layers in any way, but there is now a move towards the adoption of a standard naming convention to permit easier transfer of data between applications and users.

The British Standard advises the use of a layer naming convention based on either the CI/SfB Table 1 elemental classification, or the NBS Common Arrangement of Work Sections format. Unfortunately it does not give a positive recommendation on which to use. AutoDesk Ltd and the AutoCad™ users group have, therefore, developed a separate layer naming convention (figure 5.23), which divides the layer name into six fields.

(1) A single alpha character to identify the construction discipline assigned to that layer. For example, B indicates a building surveyor.

(2) The CI/SfB code is entered in the second field. This is a three digit field, with three levels of detail relating to CI/SfB Table 1 elemental coding. For example, level two primary elements would be represented by 200, and 210 would indicate external walls. The outer leaf of a cavity wall would be represented by 212. Within this field there are also other codes, not CI/SfB related, such as for survey grids (035). There are also some unused codes for future development.

(3) The third field is used to accept an alpha code indicating the type of data. For example, attributes, which are described below, would be represented by 'A', dimensions by 'D' and so on.

(4) Field four was originally included to relate the user defined convention to the BS layer naming convention, but in practice has not been used and will therefore be withdrawn.

(5) The fifth field specifies what level in the building the information on that layer relates to.

(6) The sixth and final field allows the user to add other information, or permits a user defined code.

B	**212**	**A**	**X**	**$15**	**GBX**
field 1	field 2	field 3	field 4	field 5	field 6/7
discipline	CI/SfB code	type	boundary	level	user codes
building surveyor	outer leaf cavity wall	attribute		level 15	inter. code (optional)

Figure 5.23 AutoCad™ layer naming convention.

A layer system, with systematic coding, goes some way to providing an information management tool, but does not fully provide the means of integrating condition data into the model.

An integrated computer model

Within AutoCad™, the user can enter non-graphic information, called attributes, alongside the relevant drawing entity. These can be visible or invisible, and originally could only be associated with blocks, which are sets of entities compounded together to form objects. For example, drawing entities such as lines are formed into a block to represent a door. These blocks might be standard blocks, or user defined. However, further system development has introduced the concept of extended attribute data, permitting it to be referenced to any entity.

Software packages such as AutoCad™ are, in themselves, very powerful data managers, and have the ability to present information in other than graphical form. For example, there are schedule generation facilities, which provide for the extraction of attribute data into an organised form. This data can also be extracted into a disk file, for processing by a range of external data management packages.

A number of facilities management groups[8,9], have begun to exploit the potential that exists, and develop fully integrated systems, based on a full building model. Archibus™, for example works from an Auto-Cad™ platform, utilising entity attributes. This sophisticated FM package incorporates a maintenance management component, based on a graphically linked database, which operates on a similar basis to the FrontLine™ system described above. The system also has powerful facilities for interfacing with external databases and spreadsheets, and also permits the incorporation of scanning techniques, which provide a major benefit in terms of entering data to construct the building model. It should be emphasised that such software is comprehensive in nature, permitting the execution of the full range of FM activities, such as space planning and utilisation. Figure 5.24 illustrates the basic structure of another AutoCad™ based package, AutoFM™, developed by Decision Graphics[10].

The major obstacle to progress at the present time is a failure by most design teams to use CAD in the design and commissioning of buildings. It is at the design stage that the formative model is best constructed. Because the CAD model is rarely available to the building manager, the setting up of such a model is an expensive undertaking, and in extreme cases full measured surveys, as well as condition surveys may be necessary.

Information sources

Sources of information for maintenance work can be divided into internally and externally generated material. Maintenance data, and information, can also be separated into that which is of a general nature, and that which is specific to a particular building or task. The former is

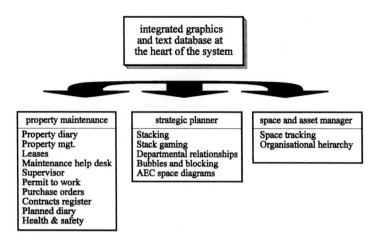

Stacking refers to a sophisticated space optimisation facility working in three dimensions.

Space tracking provides the facility for managing spaces, grouped organisationally or by other criteria, and the assets, facilities, equipment and people within them.

Figure 5.24 AutoFM™ structure.

most likely to emanate from an external source, whereas the latter will probably be internally generated. Large estate owners, however, may generate data having more general application, either internally or externally. The maintenance of good external data sources also requires a flow of data from individual organisations to produce a sensible data bank.

There are three major issues to be considered.

(1)   Information channels may be formal or informal, and the importance of the latter must not be underestimated. The major problem, however, is one of harnessing and organising this material.

(2)   At a national level there is a proliferation of material, and this presents a major organisational and analysis task. Transfer of the information to the most appropriate point of use is a well known problem.

(3)   Raw data, in whichever direction it flows (internal to external or vice versa), requires manipulation into meaningful information, and this has to be executed with a great deal of caution. Information of a general nature often only represents an average stance.

This means that, on a national basis, the industry is confronted with a critical information management task.

Internally generated information

The flow and storage of data throughout the life of a building, from inception into occupation, has already been discussed. The main vehicle for collection of data is the building data store, or building performance model. Such a facility will always exist, in a form ranging from the primitive to the technologically sophisticated.

The specific items of data and information generated in this way are illustrated in figure 5.25, where the categorisation used is by point of generation. This is essentially a continuous process and this illustration represents a simplified model.

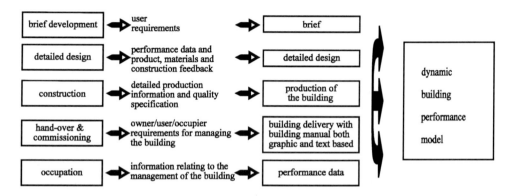

Figure 5.25 Information inputs and outputs through the building life cycle.

Information from external sources

The following external data sources are of significance.

(1) Specific government departments have responsibility for managing the buildings which house their various functions, and these provide data and information on the performance of their buildings in use. They will, for example, give design guides and recommendations on maintenance procedures. In addition there is a limited amount of cost-in-use data available, and they may transfer some of this data to other collecting agencies.

The DHSS, in maintaining hospital buildings, provided design guidance and useful information with respect to maintenance and cleaning. Much direct responsibility was handed down to area health authorities, and now to self-governing hospital trusts. Increasingly, therefore, the responsibility for generating maintenance data will reside at local level and it is too early to assess the effect this may have on dissemination of information.

The Department for Education and Employment also has an input to standards required for school maintenance, although on the ground authority rests with the LEA, and, to some extent, opting out schools.

(2) Local authorities generate information on many buildings for which they are responsible. Although little of this information is published, it finds its way to other agencies. A major contribution is with respect to housing maintenance, and there are a number of formal and informal groupings that provide valuable collection and dissemination points.

(3) A significant portion of, what may be termed social housing now rests in the hands of housing associations, who are overseen by the Housing Corporation. The Housing Corporation, through the financial control it exerts, and as an advice centre, provides important data.

The National Federation of Housing Associations (NFHA) also acts as a useful co-ordinating body, and research sponsor. Much of their research work is, however, currently directed towards social aspects of housing.

(4) The Department of the Environment publishes important material relating to all aspects of the building and construction industry, often in conjunction with the Government Statistic Service. The Property Services Agency (PSA) was set up within the DoE in 1972 to advise government departments on aspects of property management, construction and maintenance, and to have more direct responsibility for a large part of the public building stock, including military establishments and so-called Crown Buildings. This effectively made the PSA a major estate owner, enabling it to become an important information resource. It facilitated, for example, the provision of a library and the publication of many important documents, including standard specifications and schedules of rates of specific benefit for maintenance work. The dismantling of the PSA may well, therefore, have far-reaching implications for building maintenance information.

(5) The National Audit Office is a government watchdog, charged with appraising the effectiveness of a range of public sector organisations, with the Audit Commission having specific responsibility for local authorities in England and Wales. Whilst their reports tend to represent overviews, they are of great value in terms of the principles they outline and the guidelines they lay down.

(6) The Building Research Establishment is an important source of information for all sectors of the industry.

(7) The Chartered Institute of Public Finance and Accounting pub-

lishes annual statistics for the maintenance costs of public housing, and advises on accounting procedures that may be used for this type of work.

(8)   The British Standards Institute publications, whilst of general applicability, provide important maintenance specific recommendations. The time taken in drafting standards means that they do at times tend to lag behind current practice.

(9)   The British Board of Agrément was set up in 1966 to provide technical assessment of new products. Their certificates of approval include an assessment of the probable performance of the product in its intended use and, in appropriate cases, the maintenance requirements and probable life. Although of particular use for maintenance purposes, they only cover a limited range of products.

(10)   Manufacturers' information is produced in large quantities, although its quality, in information terms, is extremely variable.

(11)   Trade associations, set up to promote the proper use of specific types of materials and products, provide useful maintenance information, and will often give advice on specific problems.

(12)   Professional bodies have maintenance interest groups, which produce a range of valuable publications. One of the most important of these is Building Maintenance Information (BMI), an RICS company. This replaced the former Building Maintenance Cost Information Service.

(13)   There are a number of other bodies that either directly, or indirectly, support maintenance activity. These include the Joint Contracts Tribunal, the Association of Facilities Managers and the British Institute of Management.

Most of this external information will tend to be of general or background significance, and its interpretation may be difficult. A major issue is the means of accessing this information. This is now increasingly becoming automated. It also raises the question of a suitable means of classifying information.

Classification and coding of maintenance information

General information

Most general information will be generated externally, and will be classified using industry standard methods. The two most relevant classification systems are:

❑ Universal decimal system
❑ The CI/SfB system

The former of these is rather rigidly hierarchical and, whilst useful for accessing published data, is of little use as an aid to data manipulation.

CI/SfB was set up initially for new-build work, as a response to the information explosion. It may become an important component of a more specifically designed information system for maintenance activity.

Also of relevance here are the Common Arrangement proposals of the Committee for Co-ordinated Project Information (CCPI)[11].

Specific maintenance information

Very specific maintenance information does not lend itself to classification using either of these methods alone. However, internally generated reports, on the performance of a particular product, can easily be given a CI/SfB classification. The major classification problems, and hence coding difficulties, stem from the characteristics of maintenance operations.

Not only is maintenance information extremely fluid, in the sense that each stage constantly updates and passes on information, but there are a number of characteristic items of information associated with a maintenance operation:

❑ Location of the operation
❑ Building element or group of elements
❑ Function of element/location
❑ Reason for the operation
❑ A description of the operation
❑ Who did it and when
❑ Magnitude or seriousness
❑ Frequency of occurrence
❑ Resources expended and time taken
❑ Budget and actual cost
❑ Source of funding

There are any number of combinations of these that can be used to order and group information, all of which require a coding system of one sort or another. In a sense, these items represent the raw data of maintenance activity, and their grouping, sorting and coding is a fundamental information management issue. This means that the degree of coding sophistication should be determined by properly defined management information requirements. If independent management functions are considered in isolation, then a simple approach would suffice.

For everyday operational management, sort criteria, based simply on location and work gang, may provide an appropriate basis. Lee[12] proposes a combination of location and element as a basis (figure 5.26). This approach has some merit also, in that it is consistent with the way in which inspections and condition surveys might be carried out, which opens up the opportunity for an integrated electronic data capture and storage approach.

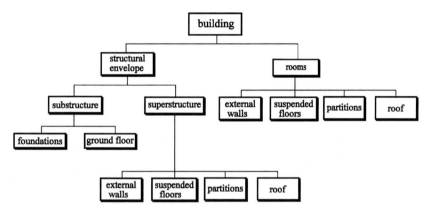

Figure 5.26 Information classification based on location and fabric element.

The structure, outlined in figure 5.26, can be supplemented by a code that describes the function of a space. On the other hand, for financial management purposes, it may be necessary to group items in terms of cost centres, or accounting periods.

For design feedback purposes, elemental or functional criteria may be the most appropriate. There are, for example, a number of published information sources in the form of elemental cost analyses of buildings in use. Lee[13] identifies four structural approaches:

❑ Those paralleling the ones used for new-build purposes
❑ Those designed to facilitate the collection of data[14,15]
❑ Those that are an extension of accounting procedures[16]
❑ Those designed to permit cost comparisons of maintenance expenditure incurred by different organisations

In designing an integrated management information system, it is apparent that there may be a number of conflicts. This highlights the usefulness of relational databases, particularly when management requirements dictate the necessity for an analytical capability on a large estate.

This requires the use of a coding regime at the data source, i.e. execution of work. The coding of items themselves, however, has to emanate from management. The data collection demands at the workface should always be strictly limited. The prime requirement at this point will probably be the collection of labour data, and perhaps material usage. For a number of reasons, this data needs to be treated with caution.

The most appropriate methodology, at the present state of the art, appears to be a sort code approach. An example is shown in figure 5.27. This is not incompatible with the layer naming strategy associated with CAD systems. Obviously, however, such an approach becomes increasingly cumbersome with increasing numbers of fields.

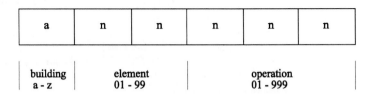

Figure 5.27 Use of sort code for coding maintenance information.

Other approaches involve the development of a standard maintenance operation phraseology, which may have benefits in terms of describing actual operations. To some extent, a standard phraseology does exist in standard schedules of rates and specification clauses. Many computerised maintenance management systems will incorporate, or link with, such documents, particularly for costing and budgeting purposes.

References

(1) Tricker, R.T. (1982) *Effective Information Management*. Beaumont Executive Press, London.
(2) Edwards, C., Ward, J. & Bytheway, A. (1991) *The Essence of Information Systems*. Prentice-Hall, London.
(3) Edwards, C., Ward, J. & Bytheway, A. (1991) *The Essence of Information Systems*. Prentice-Hall, London.
(4) Lee, R. (1987) *Building Maintenance Management*. Blackwell Science Ltd, Oxford.
(5) Klammt, F. (1990) *Lessons Learnt from Installing Maintenance Management Systems – Facilities 90: Proceedings of the Computer Aided Facility Management and High Technology Systems Conference*. Atlanta, Georgia.
(6) McKellar, W. (1990) *Nestle Approach – Proceedings of the 13th National Maintenance Management Show*. London.

(7)   Lovejoy, D. (1990) Getting the most out of CAD, *Facilities*, vol. 18(12), December. London.

(8)   Donaldson, I. (1991) The scope of computer aided facilities management: part 1, *Facilities*, vol. 9(11). London.

(9)   Donaldson, I. (1992) The scope of computer aided facilities management: part 2, *Facilities*, vol. 9(12). London.

(10) Atkinson, P. (1994) *Computer Aided Facilities Management: Why,? How,? When,?* Undergraduate Thesis, De Montfort University, July.

(11) Co-ordinating Committee for Project Information (1987) *Co-ordinated Project Information for Building Works – a guide with examples.* CCPI, London.

(12) Lee, R. (1987) *Building Maintenance Management.* Blackwell Science Ltd, Oxford.

(13) Lee, R. (1987) *Building Maintenance Management.* Blackwell Science Ltd, Oxford.

(14) Bushell, R.J. (1970) Building planned maintenance, *Building Maintenance*, September. London.

(15) Department of the Environment (1970) *Practice in Property Management – Research and Development Bulletin.* HMSO, London.

(16) Jarmann, M.V. (1967) Selling Maintenance to Management, *Proceedings of Conference on Profitable Maintenance.* London.

Chapter 6
Maintenance Planning

Introduction to planning principles

The process of planning for maintenance work has much in common with the planning of any construction activity. Therefore, the basic principles of planning should be firmly understood before considering maintenance planning specifically.

Essentially planning must be seen as a thought process. Whatever activity is engaged in, whether consciously or sub-conscientiously, some plan is formulated mentally. In many of these cases there will be no formal commitment on paper, but an intellectual process will have been engaged in to get from point A to point B, or to make product X.

As the nature of the product or activity becomes more complex a point is reached where it becomes necessary to commit some, or all, of this plan to paper and a formal programme is produced. At a simple level, this may only involve writing dates into a diary whilst, at a more advanced level, the use of a powerful computer based management technique may be necessary.

The point at which the transformation from a simple representation to a more sophisticated one occurs is imprecise, and dependent on a great number of factors, not all of which are necessarily related to the complexity of the task being planned. The use of sophisticated planning techniques may appear as something rather clever, but in reality they are only as good as the thought processes underlying them.

Objectives of planning

Planning, as an intellectual process, permeates all activities in one form or another, always with some objective in mind, whether or not this is overtly stated. The clear identification of objectives is an essential pre-requisite of the whole process, but particularly prior to the committal of a plan to the formal programming process.

186

In the construction industry, planning has all too often been afforded insufficient credence. In many cases this is because not enough attention is given to the purposes for which a plan is required, leading to a failure to produce programmes that are consistent with the planning objectives. This tends either to bring the planning process into disrepute, or to the setting up of an intensely bureaucratic management regime.

Essentially, a planning team is concerned with the provision of management information, and the purposes for which this information is required will be the over-riding factor in determining the most appropriate means of producing it. A systematic review of needs leads to the development of a number of key descriptors, which summarise the essential requirements of a planning system.

A major function of project management is that of control. To control something implies measurement, comparison with a benchmark, drawing conclusions and taking appropriate action (figure 6.1). An objective of project planning, therefore, may be to lay down a formal benchmark in the form of a programme, to act as part of a control system.

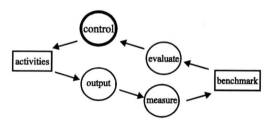

Figure 6.1 Simple control system.

A number of aspects in a project may require control. The most obvious of these is time, and it is with this that construction programmes are normally concerned. However, these programmes are not exclusively time controlling tools, and they may need to be used for a number of other purposes. There will, for example, be an interest in controlling the rate at which money is spent, so that cash flow forecasts and budgets will need to be produced. This is, of course, of major importance in maintenance management. In order to produce a financial plan a temporal plan is required, so these two are intrinsically linked.

There are other aspects of a project which need to be controlled such as quality. Quality benchmarks are set through drawings and specifications, and these can be viewed as representing a model of the project. In the same way a programme can be viewed as a temporal model of a project.

In order to be successful, a programme needs to provide as accurate a predictive temporal model of the project as possible, against which real progress can be measured and assessed, so that the *status* of the project can be determined. Within a proper control system the programme must be able to extrapolate forward from measured progress and thus be *predictive*.

The process is essentially an on-going one throughout the project and, as a control system implies the taking of action, it is useful if the planning system also acts as a *diagnostic* tool. For example, how was the current status achieved i.e. the programme provides a *historical record* that can be used to carry out a diagnosis of project difficulties.

If the programme is to be successful in providing a temporal control model, it must be *appropriate* for management purposes, and should include information that is both *relevant* and *realistic*.

The planning process does not end with the commencement of a project, so a good system should be *dynamic*, in the sense that it moves with the project. Too often construction programmes are produced under a very rigid regime with a quite unreasonable reluctance to revise them as circumstances change. If the programme is a genuinely dynamic model it has to be capable of responding to changing events and be *flexible*.

As a management information service, the planning system has to act as an aid to decision making and permit management to ask questions and evaluate the consequences of alternative actions. In order to provide this service planning systems should be *interrogative* and *interactive*. Overlying all of these requirements is the need for a programme to present its information in a clear and concise way and thus be *communicative*.

The achievement of these requirements, represented by the highlighted keywords, depends not only on the skill of the planning team but also on choosing the most appropriate planning techniques, and for this reason it is important that the components of a programme are properly understood.

The components of a programme

In the discussions below the term project is used, and this should be viewed in a very broad context that includes the planning of maintenance work. Four distinct but inter-related components of any plan can be identified.

(1) A number of discrete activities or tasks
Whether viewed as a thought process, or as a formal plan, any project

will comprise a series of steps which have to be taken in order to move towards the attainment of the end goal. In a programme these will be represented by a series of what are normally referred to as activities. The preparation of a programme, therefore, requires the breaking down of a project which is in reality a continuum, into discrete activities. This is often considered a fairly simple, if laborious, exercise to be dealt with as easily and quickly as possible in order to move on to the 'real' task of playing with logic diagrams and computer displays. In actual fact it is at this initial stage that the real skill of the planner is tested most severely, as there are a number of questions that have to be resolved.

In the first instance, projects do not necessarily divide themselves into discrete self-contained activities, so that divisions need to be considered carefully. Too often insufficient thought is given to this process, and decisions are made in a somewhat arbitrary way. On the average construction project, for a new building, a series of activities can be readily identified, and the assumption is then made that these can be linked together into an orderly sequence that represents the construction process. In reality this is a simplistic, and an almost completely erroneous view, as experience shows that activities overlap, are discontinuous, and more often than not do not proceed in an orderly sequence. This leads to a quite unrealistic assumption about the nature of the planning process, and it is essential that a planner be aware of the limitations of what he produces.

The way in which the project is broken down is, therefore, of great importance to the success or otherwise of the planning process. Initially, an answer is required to the fundamental question as to how many activities are necessary, and this will determine the level of detail of the programme produced.

To answer this question two important issues have to be addressed:

❑ The purpose for which the programme is being produced and the management objectives attached to it
❑ The quality of data, in terms of detailed knowledge of the project, available at the time the programme is being produced

Programmes are produced for many purposes, and it is unlikely that one detailed programme will serve every purpose. For example, a programme may have insufficient detail because the programme is not broken down into enough activities whilst, in other cases, a fully detailed programme may be inappropriate for a senior manager who only requires an overview.

With respect to data quality, the uncertain nature of construction activity has to be recognised, and it must be accepted that, in the early stage of a construction project, only a limited amount of detailed information about a building is likely to be available. At this point any attempt to plan a project in minute detail, as well as being inappropriate, will also be futile because it will inevitably involve making assumptions that will be overtaken by events, sometimes before the programme has been circulated.

The level of detail of the programme must therefore be consistent with the quality of information available at the time of its production and appropriate to the objectives for which it is being produced. As an illustrative example, figures 6.2, 6.3 and 6.4 show, in simple bar chart format, three programmes that might be produced by a contractor for the same project, at various stages in its life, for different purposes.

The major challenge associated with planning maintenance work is the need to deal with a very large number of, often small, activities of a jobbing nature. This will require maintenance to be considered at a number of different levels, ranging from the operational to the strategic, where the level of detail in terms of management information will vary considerably.

(2) Timescales/activity duration

Having broken down the project into a series of activities, the next requirement is to place a timescale against each of them. In the construction of new work there are several schools of thought.

❑ Information from bills of quantities may be used to work out activity duration, but there are numerous reasons why this is not a good practice
❑ Work study data, if it is available, may be used to work out durations, although this is becoming increasingly infrequent for new construction work
❑ A mixture of assessment and experience may be used which, given a healthy degree of realism, may be as good a method as any other

The nature of new-build work, from the contracting point of view, has changed so rapidly that it is now virtually the norm for the main contractor to subcontract the whole job in various packages, so that project planning and control is less to do with in-house productivity and more to do with co-ordinating a group of separate work forces. Within this context the determination of activity durations tends to become somewhat politically influenced.

Maintenance work clearly differs somewhat from new work and, to some extent, a good planning system may generate its own data in order to enable sensible predictions about activity duration to be made. A

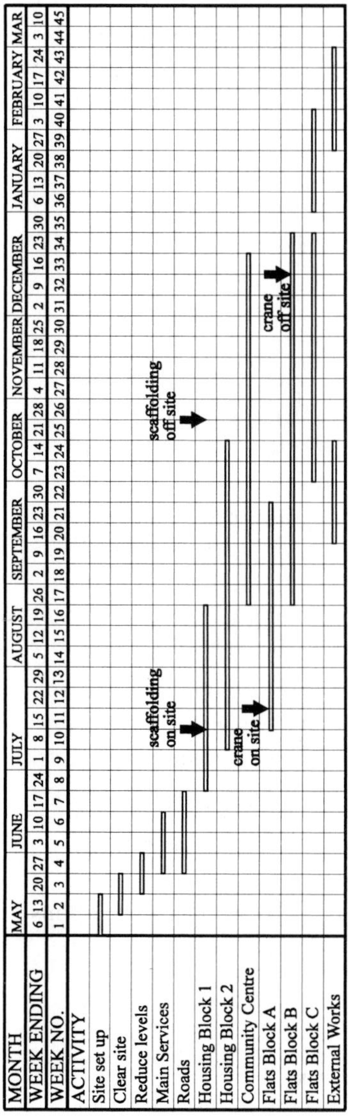

Figure 6.2
Tender programme.

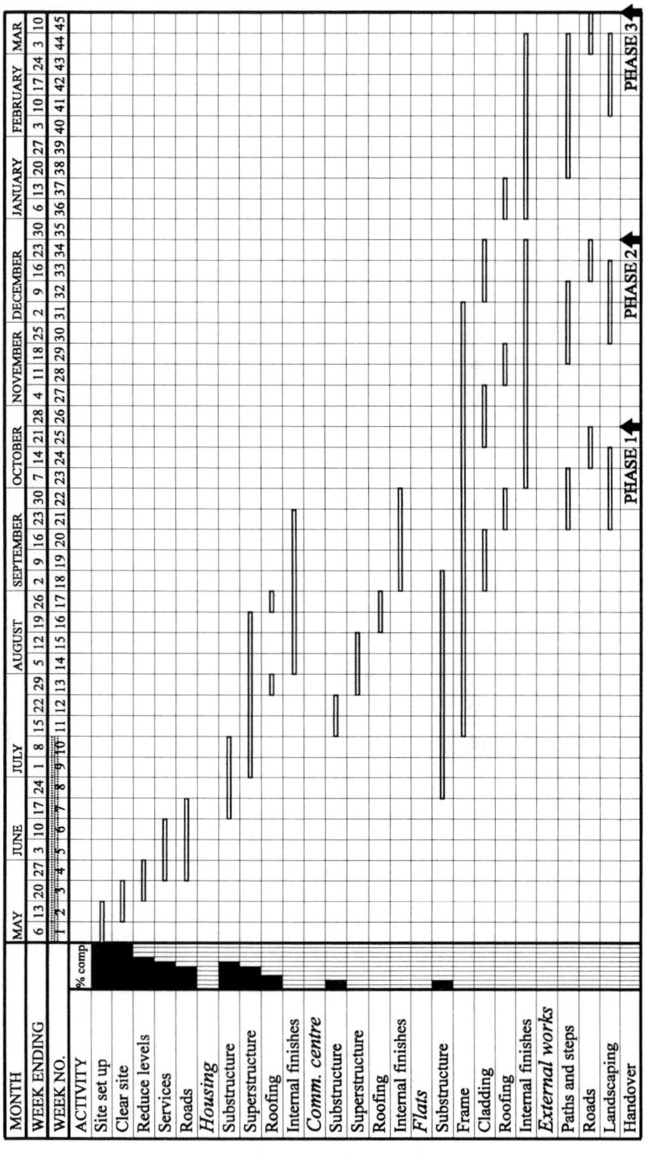

Figure 6.3
Master programme.

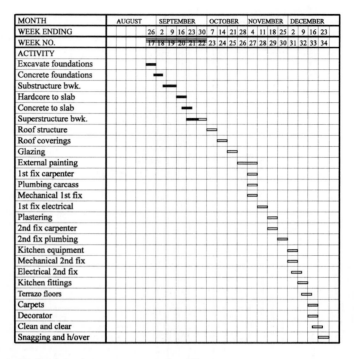

Figure 6.4 Section programme – community centre.

much more difficult problem in respect of maintenance work concerns not the individual task duration, but the prediction of appropriate cycles for regular activities.

(3) Sequence of activities
A large part of the mental process of planning is related to the ordering and sequencing of activities in a logical manner. The problem here is that the translation of a mental process, which might well view a project as a continuum, into a formal programme statement tends to require the breaking down of the project into a series of discrete activities. This is fraught with the difficulties identified above, and, in order for a plan to be truly dynamic, enabling it to be used for diagnostic purposes, these discrete activities should be formally linked in some way.

If the programme is to operate properly, within an effective control system, then an accurate picture of the current state of a project is essential. To derive this it is necessary to know the effect on a project of the state of progress of a number of activities. In practical terms this provides something of a dilemma.

A range of planning techniques are used for building work, ranging from the simple bar chart to sophisticated network systems. The former has the advantage of excellent visual presentation but, in its simple form, activities are not linked, so that it is difficult to assess the overall position of the project. This is demonstrated in figure 6.5.

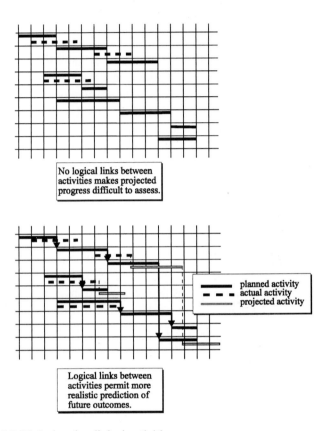

Figure 6.5 Linked and unlinked activities.

On the other hand, a network, which satisfies this requirement, is, in its purest form, a poor visual communicator. The cheaper and faster computer hardware and software programs, now available for construction work, overcome this deficiency by having clear graphical representations, thus permitting the potential benefits of networks to be exploited.

There is no doubt that the use of networks do have many benefits, not the least of which is the discipline they impose on the thought process of planning. Most maintenance work does not, however, justify their use, although there is a need, in some cases, to consider the inter-relationship of activities.

In general the over-riding imperative in maintenance planning is to communicate information clearly, and this will determine the choice of methodology.

(4) Recording of progress

The final ingredient, in a plan or programme, is a method of recording actual progress against that which was planned. Two ways of achieving this are illustrated in figures 6.3 and 6.4.

Of great significance, in respect of recording progress and improving presentation, has been the development of user friendly, readily available computer software, that provides the means to carry out analytical exercises with much more freedom. The major issue therefore, given the characteristics of maintenance work, is that of data management and this is not only a question of choosing the correct database or software package, but also of structuring data in the most appropriate fashion.

Maintenance planning

Scope of maintenance planning

Chapter 4 identified several groups of operations that may contribute to a maintenance workload and introduced the concept of planned maintenance. However, there are a number of aspects of maintenance that require planning, which may not necessarily be part of a formal planned maintenance programme. For example, it may have been decided to institute a programme of planned inspections to verify that statutory requirements are being fulfilled, or considered prudent to operate a planned replacement policy, as part of a preventive maintenance programme. This may operate separately from an on-going planned maintenance programme.

Both these will require the maintenance management team to engage in an analytical planning process, to determine appropriate cycles for inspections and replacements, concurrently with suitable presentation methods. The process requires a scientific approach, demands good data, and demonstrates that the real scope of maintenance planning is much wider than simply plotting a series of cyclic activities on to a bar chart.

Planned versus unplanned maintenance

Within any maintenance organisation there will be planned and unplanned work. The balance between the two will vary, depending on the nature of the organisation and its attitude to building maintenance. There will be an optimum balance between them which is right for the organisation concerned. A low level of planned maintenance in an organisation does not necessarily reflect a poor attitude, as it may be appropriate for the given situation.

It is quite possible to envisage a scenario where the introduction of a sophisticated planned system is not justifiable. For example, the owner of an estate consisting of one relatively simple building may choose to carry out all maintenance on demand, and plan only relatively obvious items, such as a redecoration every four years. Even the latter may be on a rather *ad hoc* basis. This closely mirrors the approach of the owner/ occupier of a dwelling house, and is an inevitable consequence of work which is characterised by a large number of relatively small, low level operations and a small number of larger ones. The latter are more likely to be foreseeable ones, and hence planned for. They are likely to fall into two categories.

❑ A regular on-going requirement to perform certain operations, such as decoration. These tasks will tend to be cyclical in nature and, in theory at least, quite conveniently form part of a rolling programme.
❑ Major renewal or repair projects which, from time-to-time become necessary. For example, there may be a programme instituted by a housing association to replace all flat roof coverings over a fixed time period.

Some of these larger exercises fall into the category of what may be termed preventive maintenance, and need to have been subjected to a rigorous decision making process. For example, a decision to replace flat roof coverings ahead of failure is a preventive measure. In reaching this decision, account would have been taken of the disruption and possible consequential damage of not replacing until failure had occurred.

A large number of very small jobs causes complications, in that it is not easy to make a case for a planned replacement or repair as part of a preventive policy for every item that may go wrong or wear out. An example of this might be the replacement of ball valves.

There may well be circumstances when a planned replacement policy is viable, but in normal work this is unlikely. Thus, a large number of items might be considered as unpredictable and difficult to plan, and in a very small estate this may prove to be quite satisfactory, provided the response time to a maintenance request is short. However, in a large estate it can result in both a poor service and an inefficient one. A policy of planned inspections of property, on a regular basis, provides the means by which a proportion of, otherwise unpredictable, items may reasonably be anticipated and placed into a planned programme.

The advantages in a large estate of maximising planned maintenance can best be visualised by considering what happens if little or no planning takes place. The maintenance workload will be characterised by a large number of small jobs, mainly of low value, distributed over a wide area, and occurring in an unpredictable pattern. Not only will these jobs be disproportionately expensive to execute, but there may also be an unacceptably long response time. To provide a highly responsive service, in such circumstances, there would have to be a pool of labour ready to respond on demand.

This is clearly expensive, as illustrated diagramatically in figure 6.6. The consequences would be a high proportion of non-productive time, probably high administration costs, and a work programme that is extremely difficult to control.

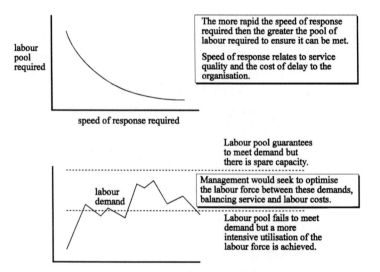

Figure 6.6 Effect of response times on labour pool.

Optimising planned maintenance requires a substantial amount of decision making to take place with respect to work cycles and replacement. If executed properly, planned maintenance work will be carried out in a more efficient way, as items can be grouped and the advantages of economies of scale can be realised. However, even with a rigorous analytical regime and precise inspection policy, there will be an element of unpredictability. The problem that then occurs is that the incidence of unpredicted demand may take resources away from the planned programme. This will have two consequences. In the first instance, a decision will have to be made with respect to the speed of response that is required, and the implications this may have on resources. Secondly, there will be a need to decide which planned item is to be taken out of a programme to resource the unplanned one.

This situation emphasises the need for the inclusion, in the maintenance plan, of some form of priority coding, coupled with provision of a sensible contingency to allow for non-planned work. Good maintenance records and historical cost data may help provide the latter, so that the maintenance plan strikes a balance between these interacting, but conflicting, workloads, and consists of planned or scheduled work alongside a contingency system (figure 6.7).

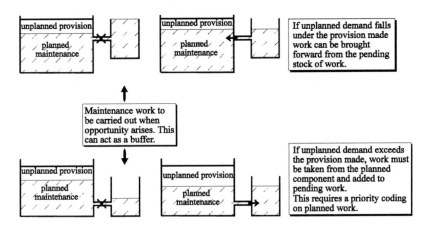

Figure 6.7 Planned work and contingency for unplanned work.

Planned inspection cycles

Planned inspections may be driven by a variety of factors, including statutory requirements. The frequency of some of these inspections may be clearly prescribed, particularly in the case of mechanical plant. These cycles may, however, be much shorter than would be dictated by fabric considerations alone and, to complicate matters even further, they may themselves all be different. For example, some items of machinery may require six-monthly inspection and others annual ones. A conscious decision may be made to inspect everything at six-monthly intervals, as the additional cost may be minimal in terms of the potential to identify problems before they occur. The appropriate frequency of inspection of the fabric of the building is by no means easy to optimise, and may, or may not, coincide with the foregoing. It is likely, therefore, that there will be a number of different inspections required at a variety of intervals.

Once the periods between inspections have been identified, it should be possible to co-ordinate them. However, the information collected during an inspection may suggest revised future inspection cycles. For example, it may be noted that an item is deteriorating to the extent that its replacement is becoming likely, but with some uncertainty attached to the diagnosis. This might suggest a future inspection on a shorter cycle than had hitherto been the norm. Practical considerations will also often dominate the timing of an inspection, such as the case when an inspection is carried out as part of a physical maintenance operation.

It is possible, if appropriate data exists, to build a mathematical model to determine an optimum inspection cycle for one item. To carry this out for every item included in an inspection routine would clearly be a major exercise, and not normally justifiable. It is, however, instructive to consider how such a model might be constructed, as it demonstrates the factors that need to be considered.

There is a presumption that a regular inspection policy will reduce emergency repair work. The first benefit that accrues, therefore, is to reduce the cost of a repair or replacement by executing it within a planned programme, rather than as an emergency item. Further benefits proceed from this, in that there will be a saving in costs associated with disruption to building usage caused by element failures. However, on the negative side the inspection has a cost attached to it, and may itself impose disruption. An analysis of this type has been carried out, using a mathematical model, by James and Green[1]. Their model, which is based on gully inspection, takes into account the following variables:

❑ The average number of call outs per annum
❑ The frequency of planned inspections
❑ The average cost of an emergency repair
❑ The average cost of planned inspection
❑ The number of gullies in the inspection programme

Another variable that might be included is the consequential cost of a gully becoming blocked in the event of an unpredicted failure.

It is necessary to make an assumption about the relationship between the number of emergency call outs and the frequency of inspection. In general, in this sort of situation, an exponential relationship, as indicated in figure 6.8, is assumed. That is, the number of call outs reduces as the frequency of inspection increases, but not linearly. To carry out a full analysis, a considerable amount of statistical data is required for the

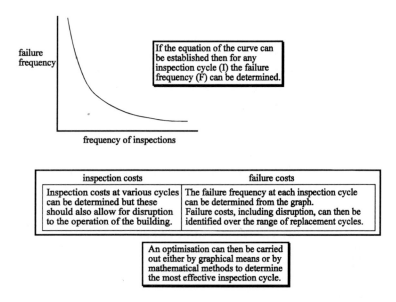

failure frequency

If the equation of the curve can be established then for any inspection cycle (I) the failure frequency (F) can be determined.

frequency of inspections

inspection costs	failure costs
Inspection costs at various cycles can be determined but these should also allow for disruption to the operation of the building.	The failure frequency at each inspection cycle can be determined from the graph. Failure costs, including disruption, can then be identified over the range of replacement cycles.

An optimisation can then be carried out either by graphical means or by mathematical methods to determine the most effective inspection cycle.

Figure 6.8 A model for optimising inspection cycles.

variables identified above, and the actual equation of the curve shown in figure 6.8 has to be determined.

Replacement decisions

Replacement decisions tend to divide into two categories, which are not mutually exclusive. In the first instance, there are replacement decisions based on the known cost of continuing to repair an item. The problem is to determine when the time for replacement has come or, better still, to predict when it will occur. There are many instances when this decision will be a professional one, based on feedback from a programme of planned inspection. For simple items, the decision may be an easy one, requiring the inspector to judge against a simple criterion.

It is also possible to take an analytical approach, and this may be justified in the case of more costly items. Provided cost data is available on historical and future maintenance costs, and the cost of replacement, then a discounting exercise will help decision making. Probably the best approach in such circumstances is to convert the future replacement cost to an annual equivalent revenue cost which, when combined with predicted maintenance costs of the new item, can be compared with the cost of maintenance if nothing is done. This technique can be adapted to make a decision now on whether to replace, or to help determine an optimum replacement date in the future.

As noted earlier, discounting techniques need to be treated with caution and depend on the availability of adequate data. Even if maintenance records are sufficiently well developed to provide basic repair costs, there are other costs and/or benefits that need to be considered. For example, a repair or replace decision needs to take into account the possibility of a complete failure that will impose disruption costs. These will probably be even more difficult to predict with accuracy.

The other approach to replacement decisions is to adopt a policy of planned replacement on a cyclical basis. For instance, a decision model is available to assist managers to decide whether or not to replace light bulbs on a cyclical basis, rather than on demand[2].

This type of technique requires data on:

❑ The failure profile of bulbs
❑ The cost of individual re-lamping, which needs to include travelling time and the cost of accessing the bulbs
❑ The cost of bulk replacement, which should enable benefits of economies of scale to be derived
❑ Consequential costs of an individual bulb failing

It is not difficult to envisage situations where a planned replacement policy may pay off. For example, in an industrial building with very high ceilings and ceiling mounted lighting fittings, access for replacement may require the use of a tower scaffold, so that the cost of the failure of one fitting can be close to the cost of a complete replacement programme.

To make such a policy feasible the decision making process will also have to determine the optimum replacement cycle. This will be very heavily dependent on having an accurate failure profile and the consequential cost of failure. Figure A1.19 in Appendix 1 and figure 6.9 illustrate how this may be approached, and show how management has to make a decision as to what represents an acceptable probability of failure, taking into account an appropriate balance between the cost of failure and the cost of a planned replacement.

Planned maintenance cycles

There will be a number of items, whose inclusion in a planned maintenance programme is unquestionable and, for these, the main requirement will be the determination of an appropriate cycle period. The obvious example is decorating frequency. There is a considerable amount of data available on decorating cycles[3], and a number of decision models have also been produced to permit analysis of the problem[4]. These should take account of:

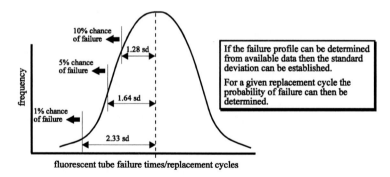

fluorescent tube failure times/replacement cycles

replacement costs	failure costs
Replacement costs at various cycles can be determined but these should also allow for disruption to the operation of the building.	The probability of failure can be determined from the frequency distribution. The probable cost of failure, including disruption, can be found for the range of replacement cycles

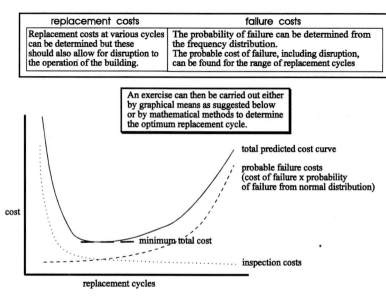

Figure 6.9 Replacement cycle model.

- Repainting costs for each cycle, bearing in mind that each redecoration may be more expensive on longer cycles due to greater deterioration
- The effect cycles have on general rates of deterioration, and the probability of accelerating complete failure of components
- Costs of disruption to productive processes, both for component failures and the decorating operation
- Aesthetic influences may have hidden costs, e.g. customer service

Some caution has to be exercised in using published data, as there are always a number of specific localised factors that need to be considered. For example, external painting cycles will be influenced by local climatic and environmental factors, and the orientation and degree of exposure of the building. This may lead to differential weathering, and force the adoption of a different cycle for some elevations. Along with specific user requirements, this may also suggest differing cycles for different buildings within the same estate.

Internal decoration may be influenced by a number of additional factors. These relate not only to the nature of internal materials, but also statutory requirements in terms of cleanliness and hygiene. The use of the building is also a factor, in terms of hygiene and cleanliness, along with the rate at which decoration deteriorates, and the aspirations of the occupiers. Internal decoration will also be disruptive, so that consequential costs need to be considered. For these reasons the use of locally derived data, if available, is to be preferred.

Forward planning for response maintenance

The striking of an appropriate balance between planned and unplanned work is, in itself, a management decision of some complexity, which can only be based on local experience. The possibility of constructing a highly sophisticated mathematical model exists, but there would be huge data requirements.

There are a number of factors that need to be considered in deciding whether work items go into the planned programme, or become part of contingency planning. These split into those relating to the technical nature of the buildings, and those likely to be determined by the organisational characteristics of the company.

In terms of technical features, clearly the rate of deterioration of components and materials is important, but this cannot be considered in isolation. In the first instance, the building's use will be important because of the way this may contribute to deterioration and, also, because certain elements may be central to the operation of the building. In pure technical terms it is the predictability of deterioration or failure that is of major importance. Items whose performance is readily predictable offer themselves very conveniently for planned inspection and repair/replacement. Other items will be more susceptible to sudden failure, and, although the execution of other maintenance work may help to identify latent potential failures, it is unlikely that this will eliminate all of them.

The important organisational characteristics relate mainly to the speed of information transfer and the service quality required.

It is almost always necessary to adopt, at least in part, a contingency approach to allow for the treatment of those items which may be equally as important as those included in a planned programme, but which may be of an unpredictable nature. This will require, as a matter of policy, the provision of contingency resources in terms of money, labour and materials in order to be able to supply a response. This is vital to safeguard the integrity of the planned programme of work, as well as to provide a service to the occupant (see figure 6.7).

The nature and size of this contingency will depend on budgeting constraints, and on the quality of service to be provided. The latter is most directly measured by a response time, that is the time which elapses between the occurrence of a fault and its repair. In practice it is sometimes taken to mean the time that has elapsed between the maintenance manager receiving notification of the fault and its repair. This approach is not to be recommended, as the speed with which a fault notification reaches the maintenance manager is an essential part of the management system. Mathematical models have also been adopted for this and, although absolute optimisation is difficult and requires accurate data, the principle is quite simple. If the required reaction time is known, and historical data is available, it becomes possible to predict the labour cost of achieving required response times.

The major problem, of course, is in deciding on a reasonable response time. This will depend on the disadvantages that will accrue whilst waiting for a repair to take place. Some of these may be quantifiable in terms of costs, but others will be much more difficult to assess. There have, for instance, been numerous studies carried out regarding the social effects of poor quality housing maintenance. One of these may be a rise in vandalism, placing increased demands on to already over-stretched maintenance budgets.

Another factor that needs to be taken into account is the nature of the defect. Some items may have greater repercussions than others and, although it might reasonably be argued that strategically important items should be part of a planned programme, it would be unrealistic to assume that this always happens. Of particular importance will be items that relate to health and safety, and those that are the subject of statutory requirements.

There are also clear cases where items of a technical nature will require a rapid response. For example, there are defects which if left untreated will become more expensive to remedy, both in terms of direct and indirect costs. A leaking roof, for instance, may not only deteriorate more quickly if left unattended, but the consequential damage will also increase. Although many cases are fairly obvious there will also be less clear ones of a marginal nature.

At the operational level there are several ways of dealing with emergency repairs, that go some way to reducing the costs and inefficiencies associated with them. In some organisations emergency repairs are logged in a systematic way. The maintenance team, engaged on their normal planned work, are required to 'pick up' emergency items at the same time as they receive their planned instructions. It has to be accepted that this may cause delays in the planned programme, although these should be allowed for, if this approach to emergency maintenance has been formally adopted as a policy.

A similar approach to emergency work is that taken in some hospital estates, where maintenance teams work in one area, for example a ward, only on designated days, at which time they will execute planned work along with any emergency items received.

It is likely that the necessity for a very rapid response tends to load more items into the contingency group in terms of execution, and may additionally put a heavy premium on planned inspections. The requirement for a rapid response capability implies that there will be larger real consequential failure costs.

Making contingency allowances for emergency work is a difficult issue. Theoretically historical data can be used and indexed for budgeting purposes, but in practice this may not prove to be very accurate. Added to this is the common problem of naturally deteriorating building fabric, which has been exacerbated by under resourced maintenance activity in previous years, leading to a backlog accumulating. This underlines the fact that historical expenditure is often no real guide to future needs.

Bearing in mind the effect that unplanned maintenance work can have on planned programmes, the tendency for backlogs to build up represents part of the vicious circle within which maintenance managers often find themselves.

Planning maintenance programmes

The aims of planned maintenance programmes are extremely diverse and, hence, many types of programme will be encountered. The application of the basic principles of planning are of paramount importance. In particular, it is essential to define the objectives of maintenance plans very accurately at the outset, to ensure their relevance, and to enable them to be realistically formulated.

These objectives may include all or a combination of the following:

❑ To help ensure that major defects are rectified and that the building fabric is maintained to a defined acceptable, safe and legally correct, standard

❏ To sustain the building condition at an acceptable level and prevent undue deterioration of the building fabric and services by preventive means

❏ To preserve the utility of the estate as an asset, and maintain its value

❏ To maintain the engineering and utility services in an optimum condition to safeguard the environmental conditions of the building, and hence its productive capacity

❏ By effective planning, to ensure that maintenance is conducted, over a number of years, in a sensible sequence which reflects a careful consideration of priorities

❏ By proper planning, to ensure that maintenance operations are carried out in the most effective way to ensure that best value for money is being obtained and the best use is being made of scarce resources

❏ To provide a tool for financial management, in particular budgetary control, and to assist maintenance managers in bidding for financial resources

❏ As part of a broader facilities management scenario, to assist management to relate programmed repairs and maintenance to other demands and alternatives, such as refurbishment, redevelopment or changes in leasing policy

The characteristics of maintenance work make accurate and comprehensive long term predictions rather difficult. It is therefore necessary to define carefully what is realistically possible, and have an explicit recognition of levels of uncertainty. Because of this all programmes will need to have built into them some flexibility to permit modification and up-dating in order to ensure their continuing relevance.

Failure to understand uncertainty and risk leads to a tendency to persist in the adoption of unreal assumptions that fail to recognise reality, and this has undoubtedly led to the misuse and abuse of modern planning techniques[5].

Maintenance has its own characteristics that further complicate the planning process.

❏ The work is characterised by a large number of small jobs, and attempts to programme individual jobs in minute detail, over more than the short term, are clearly not realistic.

❏ The widely dispersed nature of much of the work is a major factor to be taken into account when planning, as it has a major impact on efficiency and economy.

❏ Individual jobs are often simple in nature and in terms of sequencing, but there is more of a need to consider logistics, rather than detailed working methods.

❑ A large proportion of very small jobs may require the presence of a number of trades, the co-ordination of which is difficult. This makes the achievement of continuity of work for individual trades difficult.

❑ The work content of a maintenance item may be uncertain when an order or instruction is given, and the extent of a repair may only reveal itself when the building fabric is opened up.

❑ The adoption of a conscious policy for emergency items to be attended to, as part of a planned visit to a location, underlines the need for flexibility.

❑ The most carefully constructed work programmes are subject to disruption from a number of potential causes:

   ○ withdrawal of resources to deal with emergency work
   ○ climatic conditions
   ○ access problems
   ○ budgetary setbacks

❑ Emergency repairs present very specific problems, due to their unpredictable nature, often allied to the need for a rapid response, thus creating very short lead times.

In view of these difficulties, programmes are formulated at a number of different levels and each of these may be used in a different way. In general, the following categories of maintenance programme can be identified:

❑ Long term – quinquennial (five-yearly) or longer
❑ Medium term – annual
❑ Short term – monthly, weekly or even daily

In terms of management information these programmes may not all always be necessary and three levels can be considered in this respect.

(1) A programme may be required for financial management purposes only, in which case it is likely to be a long term one with, perhaps, subsidiary medium term ones. In basic terms it will comprise maintenance costings, on a building-by-building basis, perhaps with some prioritising system built in to aid decision making in the event of budgetary constraints.

(2) A further programme may be required for executive management purposes, and in this case a more detailed data input will be required in terms of a breakdown of each building's maintenance needs, probably on an elemental basis, accompanied by costings. It will also require the allocation of a time scale for operations and

costings. This type of programme is most likely to be a medium term one, but with shorter term updates or breakdowns.

(3) For operational and works monitoring purposes, shorter term programmes may be required, and these will need much more detailed information about the tasks to be performed.

Consideration of the above programmes helps to highlight several things.

❑ The various types of programme are linked and one may follow from the other through a top-down or bottom-up approach.

❑ There are no hard and fast rules about time scales for programmes; it is only necessary to follow carefully the principles, and to match these to the needs of the organisation.

❑ Typically, in any maintenance organisation, there is a maintenance planning system that represents a composite approach.

❑ For an organisation introducing planned maintenance for the first time, the cost of acquiring data is a major task, and it may well be prudent to start at (1) and build up to (3) as the data is collected.

❑ It is worth bearing in mind that the data collected will have a variety of uses, in addition to planning purposes. Indeed, it is the case with contemporary computer based planning systems that the planned maintenance programme is part of an integrated management information system.

Long term programming

The objective of long term programming is not to set down detailed task data or precise dates for individual operations. This would clearly be unrealistic, especially as the objectives of a long term programme do not require this type of detail. The application of the exception principle is of vital importance, and long term programmes should not be overloaded with spurious detail, which will only serve to confuse the real issues. To produce a long term maintenance programme may only require a systematised broad survey of the estate, such as that commonly used in the NHS[6]. In this way a picture of the overall problem can be obtained for the purposes of developing and putting forward a strategic plan of action, or to evaluate alternative strategies.

The period of study for this type of exercise will depend on the characteristics of the estate and the objectives of the organisation. There is no generally accepted industry standard. Fifteen year programmes have been proposed but, in themselves, are unlikely to be realistic. However, there is a tendency to work in five year blocks, so that a ten year programme may be produced where one level of certainty can be placed on the first five years of the programme and less on the second five.

As the programme progresses, it should be capable of revision and updating, leading to a firming up of longer term predictions as time goes by. There is now a distinct tendency for many organisations to utilise this type of rolling programme.

Longer term programmes can be produced for a variety of purposes.

(1) The determination of the expenditure required for maintenance over a period of time, in order to put and keep the building stock in an acceptable condition. Where there is a maintenance backlog problem, it is useful to separate the backlog work from future requirements when drawing up the programme. This separation will often emerge in any case, if a prioritisation system is incorporated.

(2) Possibly linked to (1), or stemming from it, is a long term programme that can be used to plan expenditure streams in the most effective way according to circumstances. An obvious example is where expenditure has to be planned within the constraints of available finance. This again encourages the use of a priority coding system.

(3) Major repair and/or renewal proposals require careful forward planning, to be consistent with the financial resources available, and to ensure that their timing has the minimum disruptive effect on the organisation. Major repair programmes are often a special case, in that they are more likely to be the result of a rigorous analysis of a specific problem, and usually lend themselves to a more precise definition of both the cost and the work content. There is often some flexibility in the planning of major repair programmes, in that they rarely contain emergency items. This fact can be very useful, as it may permit an adjustment to a programme where, perhaps, some major unpredicted failure occurs and swallows scarce resources in one financial year. An example of this is illustrated in figure 6.10.

(4) Long term forward planning is useful for predicting and budgeting for financial resources. However, it has to be borne in mind that other resources, particularly labour and materials, have to be available, broadly at the right time. Whilst this can be achieved to some extent by long term planning, detailed analysis of material and labour requirements is more effectively carried out as a medium term exercise.

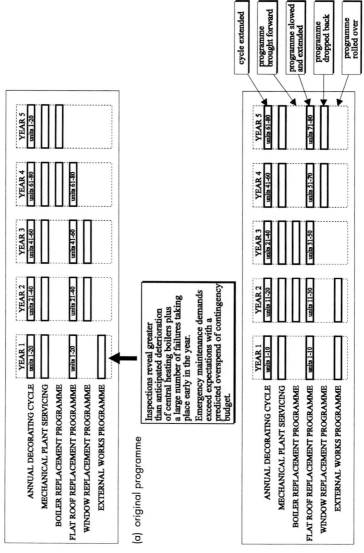

Figure 6.10 Adjustment of repair programmes over a five year period.

(5)   The broader dictates of facilities management will make use of long term maintenance plans which will need to be integrated with other longer term ventures, such as those relating to new capital projects, refurbishment, demolition and changes in the use of the building stock.

Several elements of the programme may be readily predictable, but as the timescale increases, so the level of uncertainty becomes greater. This is frequently the case with cost predictions. Figure 6.10 shows an example of part of a long term rolling programme, where a planned programme is being instituted for the first time. During the first year there are some unforeseen events, and changed conditions revealed by more detailed inspections. The programme is adjusted accordingly.

It is also possible that the need for a completely unforeseen renewal programme becomes apparent through the first year, and to accommodate it the programme would require further modification and updating. A further example of part of a long term programme, based on financial predictions, is shown in figure 6.11.

Intermediate or medium term programmes

These will normally be annually based, although updating at intermediate periods should be accepted as inevitable. The intervals at which this takes place will vary, depending on where the maintenance management system fits within the overall organisation. In many cases the reporting and control mechanisms will be driven by accounting procedures, requiring monthly reporting and review to be adopted.

At the outset, annual programmes will be set within the framework provided by longer term planning. However, at the formulation stage, there may well be specific needs that have been identified for inclusion in the programme for that year, leading to revisions in long term plans. Information will therefore flow in both directions between medium and long term programmes and this underlines their essential interdependence. At the level of the medium term timescale there will be less uncertainty, and it would logically be expected that a greater level of detail would be built into them, so that they become of much more benefit for operational purposes.

Several important uses of medium term programmes can be identified in this context.

(1)   They assist managers to determine the allocation of the annual maintenance budget, and to plan the way in which this expenditure

LONG TERM MAINTENANCE PLAN FOR MILL LANE TEACHING BLOCK
FIVE YEAR BUDGET FOR THIS BUILDING £150 000 (£142 250 ALLOCATED)
PROVISIONAL 15 YEAR EXPENDITURE CURRENTLY £325 000 (£300 750 ALLOCATED)
CONTINGENCY CENTRALLY CONTROLLED

ITEM OF WORK	PRIORITY CODE	1996	1997	1998	1999	2000	2001-2005	2006-2010
cyclical maintenance								
decoration			12 000	13 000	13 500	14 000	15 000	12 000
cleaning		13 000	13 000				50 000	50 000
external work		3 000	2 000	1 500	1 500	1 500	5 000	5 000
plant servicing		1 500	500	1 000	1 000	1 250	6 500	7 000
inspections		1 000	500	1 000	500	2 000	4 000	4 000
remedial works								
repoint verges	high	1 000						
repair paths and steps	medium		3 000					
repair damaged plaster	medium		1 000					
remedial work to internal joinery	low		2 000					
major programmes								
boiler replacement	high		20 000					
window replacement	low			5 000	5 000			
flat roof renewal	medium			5 000				
replace gutters and downpipes	low				2 000			
TOTAL EXPENDITURE		19 500	54 000	26 500	23 500	18 750	80 500	78 000

Figure 6.11 Example of part of a long term maintenance programme (based on 1996 prices).

will occur in the most effective and appropriate way, so that good short term financial planning is achieved.

(2)   From the operational point of view, money is one resource whose usage has to be controlled, and hence needs to be planned. The annual programme is the essential tool for planning the optimum usage of the resources that cost money, namely, labour and materials. If labour is in-house then use may be made of histograms, and other similar techniques, for the purpose of resource optimisation exercises.

(3)   If the work is to be executed by contractors, then the planning exercise is still just as relevant to ensure that a proper time period is available for the preparation of contract documentation and contractor selection. Where a number of term contractors are employed, their work will need to be balanced and apportioned effectively so that:

- ❑   No contractor becomes overloaded
- ❑   The right work is allocated to the right contractor on the basis of his strengths and weaknesses, wherever possible
- ❑   Geographical issues are taken into account

(4)   In terms of materials use, the annual programme should be able to provide indications of the quantities of key materials and components that will be required during the year. This will enable a buying and storage strategy to be developed, ensuring that materials and components are always available as required, and that they are obtained on the best available terms.

(5)   The annual maintenance programme should enable maintenance work to be timed as conveniently as possible, so as to minimise disruption within the client organisation and reduce real costs.

An example of parts of an annual programme are shown in figures 6.12 and 6.13, with the work broken down as follows:

- ❑   individual jobs
- ❑   division into major and routine jobs, with an allowance for emergency work
- ❑   division into that to be executed by direct labour and that by contract labour
- ❑   expenditure estimated and allocated to each job, with further subdivision into labour plant and material costs, so as to help medium term resource planning

The preparation of the annual programme can be divided into the

MILL LANE TEACHING BLOCK - ANNUAL MAINTENANCE EXPENDITURE BREAKDOWN 1997

TOTAL BUDGETED EXPENDITURE £56 000 INCLUDING PROVISION FOR UNPLANNED WORK

ITEM OF WORK	PRIORITY CODE	TOTAL EST. COST	CONTRACT	DIRECT LABOUR		
				Materials	Labour	Overheads
routine maintenance						
decoration		12 000	12 000			
cleaning		13 000		3 000	7 000	3 000
external work		2 000		300	1 500	200
plant servicing		500		100	350	50
inspections		500			450	50
individual tasks						
boiler replacement	high	20 000	20 000			
repair paths and steps	medium	3 000		700	2 000	300
repair damaged plaster	medium	1 000		200	700	100
user requests and emergencies						
		4 000				
TOTAL EXPENDITURE		56 000	32 000	4 300	12 000	3 700

Figure 6.12 Budgeted annual expenditure.

following steps, which are broadly in line with the principles outlined at the beginning of this chapter.

(1) Identification of items to be included in the programme

❏ Items brought forward from the annual programme, although they may need verification through inspection.

❏ Items identified from an inspection, which are required to be carried out in the coming year. If these items are not in the longer term programme then the implications may need to be assessed for the long term. If the long term planning has been carefully carried out, there should be a contingency item from which these items can be funded.

❏ During inspections, there may arise requests for items of work from the occupants. In some organisations it is the practice to ask occupants to nominate items prior to a regular inspection, which enables some judgements to be made in the context of budgetary and resource constraints.

❏ An allowance for unforeseen maintenance items.

❏ Routine day-to-day maintenance items, of which the greatest contributor will probably be cleaning.

MILL LANE TEACHING BLOCK - ANNUAL MAINTENANCE PLAN 1997

ITEM OF WORK	PRIORITY CODE	EST. COST	EXECUTION	JAN	FEB	MAR	APR	MAY	JUNE	JULY	AUG	SEPT	OCT	NOV	DEC
routine maintenance															
decoration		12 000	contract				4 000			2 000	4 000	2 000			
cleaning		13 000	direct lab.	1 000	1 000	1 000	1 000	1 000	1 000	1 000	1 500	1 000	1 000	1 000	1 000
external work		2 000	direct lab.									2 000			
plant servicing		500	direct lab.							500					
inspections		500	direct lab.												500
individual tasks															
boiler replacement	high	20 000	contract							5 000	10 000	5 000			
repair paths and steps	medium	3 000	direct lab.							3 000					
repair damaged plaster	medium	1 000	direct lab.				1 000								
user requests and emergencies		4 000	direct lab.	450	450	450	450	350	250	150	150	150	300	400	450

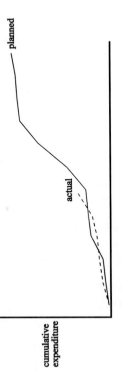

Figure 6.13 Budgeted annual expenditure and maintenance programme.

(2) Identification of work content and costs

If a good data management system exists, then feedback from previous work is the major tool to be used for this purpose. Decisions may have to be made as to which items are to be executed by direct labour, and which by contractors. A policy framework for this may already exist in the organisation.

The breakdown of the work content of the item enables appropriate material and labour inputs to be identified, and hence costs. There may, at this stage, be a number of detailed planning exercises carried out, e.g. for labour planning.

(3) Determining the sequence of work

The use of critical path networks is rarely relevant except for the biggest maintenance exercises. There will, however, often be a substantial logistics problem to solve, and thus the possible use of operational research techniques should not be excluded.

It is in this aspect of planning that computer based-systems come into their own, because of their ability to manipulate data very quickly. This enables the planner to analyse and evaluate a variety of working methods and sequences, and thus make choices that will optimise a programme. Automated data management also permits the application of a range of decision making aids.

(4) Provision of a controlling mechanism

The programme that is set up through the preceding process now becomes a model of the work of the maintenance department in the medium term, and is thus an essential component of a management control system. It is necessary for actual achievements to be compared against predicted, in order to assess progress and prompt management to take corrective action. One of the first pre-requisites here is a feedback mechanism to collect appropriate information by which progress can be measured.

In a manual planning system there are numerous ways in which progress can be recorded. These are also available in a computer based system, where data can be more easily and quickly processed. Such systems also encourage the adoption of a flexible approach to planning, as continuous updating is much more easily effected.

Short term programmes

When short term planning is considered, more detailed aspects become of importance for two reasons:

- Analysis of performance, and the provision of feedback data is the essential source of information for future planning exercises, and for this to be facilitated short term programmes require a detailed basis against which data can be collected, compared and analysed.
- For operational purposes, and day-to-day management control, it is important that all inputs to a maintenance operation are properly identified and stated.

At the beginning of each year, monthly programmes may be produced by simply breaking down the annual programme into twelve workloads. Monthly programmes produced in this way will almost certainly have limited value and, bearing in mind the principle of ensuring that planning should be relevant and realistic, caution should be exercised. However, figure 6.14 illustrates typical examples of some monthly planning exercises.

There may well be a case for taking a forward view in more detail, perhaps on a quarterly basis, to help good resource management, but, realistically, a prediction of a month ahead seems reasonable. Quarterly programmes may, nevertheless, have some benefit when a large proportion of contract labour is being used, as this is most likely to be for predictable work and the responsibility for resource management shifts to the contractor. The client, under certain circumstances and types of contract, may require the contractor to produce a programme for the work he is to execute.

When a high proportion of direct labour is being employed, short term planning may take place at a variety of levels, and the possibility of monthly programmes having to match accounting cycles, for example, has already been mentioned.

At the operational level there are likely to be weekly programmes, and sometimes daily ones. When a maintenance team collects its work allocation for a day this can be thought of as a short term programme of work. When maintenance operatives collect a works order, execute the work and complete the works order they are in fact working within a planning system.

Many of the optional contributions of the annual programme become of prime importance in the shorter term. This is particularly the case for manpower planning. Even though the annual programme and the short term programme tend to complement each other, both are necessary to make best use of the work force. In terms of operational management, the detailed programming of work depends on the extent of the direct labour contribution, and the need to ensure a smooth programme of work for it, which also reduces travelling time to a minimum.

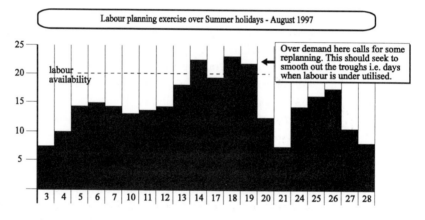

ITEM	July				August					September			
	6	13	20	27	3	10	17	24	31	7	14	21	28
Decoration		▬	▬	▬						▬	▬	▬	
Repair paths and steps		▬	▬										
External works										▬	▬		
Plant service				▬									
boiler replacement													
Prepare new boiler house		▬											
Re-route existing services			▬										
Install new boiler					▬	▬	▬						
Make connections								▬					
Commission									▬				
Builders work										▬			
Remove old boilers										▬			
Refurbish old boiler house											▬		

MILL LANE TEACHING BLOCK - major works - summer 1997 short term programme

(a) Programme of works over summer period in one building

Labour planning exercise over Summer holidays - August 1997

Over demand here calls for some replanning. This should seek to smooth out the troughs i.e. days when labour is under utilised.

labour availability

(b) Assessment of labour requirements for whole estate during August 1997

Figure 6.14 Typical short term programmes.

The short term programme, in a sense, puts into action the forward planning of the annual programme, and feedback to it will enable monitoring and control for the whole year. The three levels of planning identified above are evidently not mutually exclusive, and the boundaries between them are indistinct. All three should be viewed as integrated parts of a comprehensive planning system.

Presentation

The most familiar presentation format for a programme is the bar chart, sometimes termed a Gantt chart, after one of the early twentieth-century pioneers of management. The bar chart has the advantage of simple visual communication, and can be tailored to include many items of information. The examples shown in figures 6.2, 6.3 and 6.4, for a new-build project, are indicative of its usefulness.

The bar chart has been used for many years as a technique for new-build construction work. Its main deficiency is that on its own it does not indicate the relationship between activities. For this reason there has been, over a number of years, a growing impetus for the use of critical path networks on the grounds that they remedy this. However, in terms of presentation, they are not particularly appropriate, and to be really useful it is necessary to generate a bar chart from the network.

The advantages of networks have only begun to be realised relatively recently, with the availability of cheaper and more user-friendly hardware and software. There is no doubt that this is the case for a great number of more sophisticated planning techniques, with the added advantage of enormous flexibility in the ways in which information can be presented.

Computerised management systems are able to present information in tabular forms, as schedules, histograms and also in a graphical form. The majority of good networking packages also have the ability to export to or import data from spreadsheets and databases, so that they become part of an integrated information management system.

Networking techniques undoubtedly have their application in planning for repair and maintenance, but this is limited to larger one-off projects of a more complex nature. The majority of maintenance activity, however, is represented by a large number of small activities, many of a routine nature.

Management control

There are several components which need to be present for a control system to be effective:

- Established benchmarks or performance standards
- Measurement of performance
- Comparison of performance against that required
- Corrective action

A fifth element, that could be added to these, is the presence of a satisfactory information system that communicates, collects and analyses data, and channels it in the appropriate manner[7].

A basic question that has to be addressed, at the outset, concerns what has to be controlled, and in the context of maintenance operations this will be quality, time and money. Control methods in very simple terms can normally be categorised into:

- Those focusing on physical values of measurement such as quality, but also perhaps encompassing output performance
- Those focusing on financial values and which are intrinsically linked to time management

The normal mechanism for the latter is a budget and, in the management of maintenance, this will be an expenditure statement, and the process of control will require the monitoring of costs.

The nature of costs[8,9]

Costs can be split into four elements:

- Labour
- Materials
- Plant
- Overheads

For new-build construction activities, the separation of plant may be essential, but in many cases, when its use is of a general, rather than a directly attributable nature, plant costs are subsumed into overheads.

A direct cost is a cost that can be identified with, and allocated to, an operation. For example, materials and labour expended on a repair item are clearly direct costs, as they can be readily attributable, provided there is a proper information system. All other costs are termed indirect costs, for example, supervision or the use of small plant items not directly attributable to individual items of work.

The Institute of Cost and Management Accountants (ICMA) defines overheads as the aggregate of indirect costs. Proper financial management and control requires that overheads be accounted for in some way. In the normal maintenance department there are two categories of overhead to be considered.

In the first instance, there are the indirect costs directly associated with the maintenance department, all of which can be attributed to that department. There will also be indirect costs associated with the overall running of the company, which cannot be allocated to individual parts. In this case a method of apportionment has to be used[10].

Maintenance department budgets will, therefore, include a portion of the organisational overheads, its own attributable indirect costs and the direct cost of executing the work. Note also that, if contract labour is being used for execution, then the price paid for that service will include contractors' overheads.

Costs may vary with output, in which case they are termed variable costs. Materials, and labour expended in executing maintenance work, are obvious examples. Both direct and indirect costs may be variable.

Fixed costs do not vary with output, but will vary with the passage of time. These may also be either direct or indirect costs. It is important to appreciate that fixed costs may only remain fixed over a certain output range. For example, at a given rate of maintenance expenditure, the cost of supervision is a fixed cost. If expenditure is halved or doubled then supervision requirements will change leading to a variation in a fixed cost.

In reality many costs may fall between the two and are termed semi-variable.

Absorption and marginal costing

There are two ways of associating indirect costs with production. In absorption costing, all costs direct and indirect are charged to the unit of production, so that a unit cost has absorbed a portion of the fixed costs. If output drops, then there are fewer units produced to share these costs, so they will all need to take an extra share, i.e. unit costs increase. The converse, of course, is true and, if output rises, unit costs drop. In theory, therefore, the unit cost of one more unit is slightly less than the cost of the previous one.

In a marginal costing approach, only variable or marginal costs are charged to the process. In a commercial venture, fixed costs are written off against profit, in the period in which they occur. The cost of one more unit of production is simply the variable cost.

Both approaches have their relevant applications, but use of the most appropriate is extremely important, because of the influence it may have on management decision making. For example, a maintenance contractor may choose to take the former approach in tendering for maintenance work. A direct labour organisation, on the other hand, may prefer to operate on a marginal cost basis when deciding whether or not it can carry out an extra repair.

Standard costing

An important, and useful, concept is that of standard costing. This is a predetermined cost, derived from previous records, having regard to normal standards of efficiency, which is used as the basis for price fixing and costing. Schedules, or rates for maintenance work, or historical data, internally generated within the organisation, may be used as the basis for deriving standard costs.

Standard costs can be particularly valuable for budgeting purposes, and may also be used for incentive schemes.

Budgets and budgetary control

A budget is defined as a financial, and/or quantitative statement, prepared and approved prior to a defined period of time, of the policy to be pursued, during that period, for the purpose of attaining a given objective. It may include income, expenditure, and the employment of capital.

Budgetary control is defined by the Chartered Institute of Management Accountants (CIMA) as the establishment of budgets, relating the responsibilities of executives to the requirements of a policy, and the continuous comparison of actual with budgeted results, either to secure by individual action the objective of that policy, or to provide a basis for its revision.

In practical terms, budgets can be considered to be a planning function that produces management information. Budgetary control is, however, both a planning and an executive function, in that the term control implies action.

Within an organisation, the formulation of budgets is a complex process (figure 6.15) and requires patience and high quality management

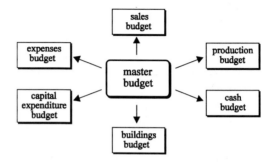

Figure 6.15 Relationship of sectional budgets to master budget.

information. Sensible costing data is obviously essential, as the budget is fundamentally a financial forecast.

Inevitably, however, the maintenance department will find itself in difficulties. Theoretically, given an up-to-date performance model of the estate, and good cost data, a maintenance manager may produce a budget, based on his medium and long term maintenance programmes. It would be naïve in the extreme, however, to assume that the budget will ultimately be driven by technical need. The maintenance manager may present a budget based on his needs, but will often be forced to accept less, so that his budget for the year is driven by financial realities.

Structure of a maintenance budget

In the same way as master budgets are sectionalised, departmental budgets are subdivided under so-called budget headings. Notwithstanding this, the departmental budget's relationship to the overall budget is important, and should be accompanied by a policy statement.

The breakdown of the maintenance budget can be carried out in a number of ways:

❏ By type of cost
❏ By type of work, e.g. a trade breakdown
❏ By building or parts of a building
❏ On an elemental basis
❏ By who will execute the work

In reality, combinations of these will be required to provide a proper control tool. However, of most significance will be a classification of the degree of importance associated with each operation.

Like any other programme, budgets should be flexible and dynamic, and there must be an acceptance that items will be dropped. Most maintenance budgets will also include a contingency item for unplannable items. On the basis that this contingency may not be totally used, the budget may also include discretionary items, perhaps ranked in terms of priority, that can be brought in. Examples of the ways in which this might operate are shown earlier.

The budget represents an incremental financial statement over a fixed period, normally a financial year. For proper control purposes, there must also be a forecast of the pattern of expenditure. If there are properly formulated maintenance programmes this information should be derived quite logically from it (figure 6.13).

A number of appendices to the budget statement may be useful. Detailed cost breakdowns and inspection reports are an example.

However, these are more properly a function of a good information system, of which the budget is a part.

Cost reporting

The maintenance information system must be designed to report actual costs against what was planned. This has to be carried out at regular intervals, normally monthly. A typical outline report is shown in figure 6.16[11].

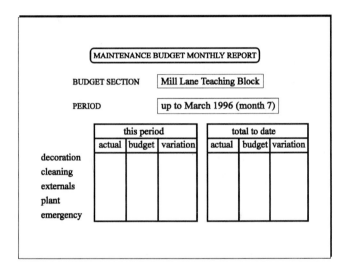

Figure 6.16 Simple budget report.

The term actual cost needs to be treated with some caution. Management information must be produced as quickly as possible, and it may be better to sacrifice accuracy in order to satisfy this need. Adjustments may, therefore, need to be made in succeeding periods. The cost variations reported may be further analysed by causes, to help management decision making. If standard costing has been used, for example, it may be useful to report variances in financial reports. Statistical techniques may be of value here, in that some variance from a standard cost would be expected. It is variances beyond what would be expected that are of importance[12].

Forecasting

Like any programme, budgets will be subjected to change and adjustment. Periodic reports should, therefore, include a predicted financial outcome, which is essentially a revised financial programme. This will need to take account of a number of different costs.

(1)   Costs that are budgeted from the original statement.
(2)   Actual costs, which may be subjected to revision in the next period.
(3)   Committed costs, which represent the consequences of irreversible decisions. The original budget may of course have included elements of this type. Also of importance, in interim reports, are costs required to complete tasks already embarked upon, which may differ from the original prediction.
(4)   Variations from budget figures in the predicted costs of future planned items, due to additional information coming to light.

The forecasting exercise will produce an up-to-date prediction of the financial outcome, and this is more useful for management control than a simple actual versus planned statement.

Corrective action

Depending on the causes of variances from the budget, and also the quality of associated management information produced, there are a number of possible scenarios.

If the budget is being, or is likely to, overrun then decisions will need to be made to take some work out of the programme or, perhaps, adopt some economies in non-programmed work. However, this only represents an immediate response, and for further management action some analysis of the cause of the problem needs to take place.

The variance may be due to:

❑   inaccurate cost forecasting, in which case remedial action needs to be taken for future forecasting exercises
❑   actual costs deviating from predictions due to inefficient working practices
❑   inefficient organisation and planning of the work
❑   variations in the scope of the work, which may be unavoidable or due to a poor inspection regime

There are clearly a large number of possible causes, but these cannot be identified properly unless the appropriate data is collected and presented. A systematic approach to the analysis of the variance is suggested in figure 6.17.

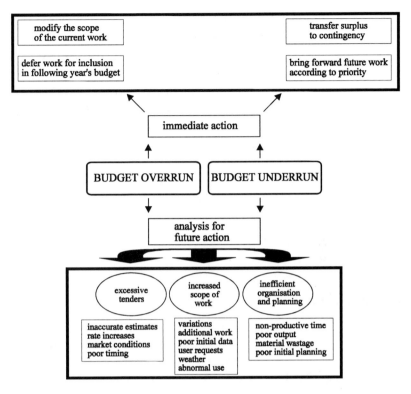

Figure 6.17 Analysis of variations from budget.

References

(1)   Gibson, E.J. (ed.) (1979) *Developments in Building Maintenance – The use of decision models in maintenance work*. Applied Science Publishers, London.

(2)   Almond, R. (1974) Cut servicing costs with systematised light replacement, *Building Maintenance, Nov/Dec*. London.

(3)   NBA Construction Consultants (1985) *Maintenance Cycles and Life Expectancies of Building Components: A guide to data and sources*. HMSO, London.

(4)   Christer, A.H. (1976) Economic cycle periods for maintenance painting, *Operational Research Quarterly*, Vol. 27, Number 1. London.

(5)   The Tavistock Institute (1966) *Interdependence and Uncertainty – Study of the Construction Industry*. London.

(6)   Sahai, V. (1987) Spedding, A. (ed.) Management of built assets – value for money. In *Building Maintenance and Economics – Transactions of the Research and Development Conference on the Management and Economics of Maintenance of Built Assets*. E. and F. Spon, London.

(7)   Cole, G.A. (1990) *Management, Theory and Practice*. DPP, London.

(8)   Davies, A. (ed.) (1971) *A First Course in Business Organisation*. Allen and Unwin, London.

(9)   Davies, D. (1990) *Finance and Accounting for Managers*. Institute of Personnel Management, London.

(10) Houlton, M.L. (1973) *An Introduction to Cost and Management Accounting*. Heinemann, London.

(11) Davies, A. (ed.) (1971) *A First Course in Business Organisation*. Allen and Unwin, London.

(12) Dyson, J.R. (1994) *Accounting for Non-Accounting Students*. Pitman, London.

Chapter 7

Maintenance Contracts

Introduction

The general legal requirements for the formation of a maintenance contract apply in the same way as they do for any other building contract. The major issue for the maintenance manager is to select the type and form of contract which is most appropriate for their purposes.

The basis of any contract is the offer by the contractor to carry out work for a sum of money, and the acceptance of that offer by the employer. Although there is no legal necessity for the contract to be in writing, it is generally accepted that this is highly desirable from the administrative point of view. It is also accepted practice that supporting documentation, termed contract documents, are prepared, to assist in the proper operation of the work covered by the contract.

There are several standard approaches to putting together contractual packages, each designed to fit a particular set of circumstances. Over the years, there has been a proliferation of standard forms, and much debate and research has taken place concerning the appropriateness of each for new-build, repair, maintenance and refurbishment work.

The main differences between the various types of contract available relate to the methods of evaluating work, and the degree of financial risk to be borne by the parties. This is reflected in differing arrangements for pricing, valuing, paying for the work, and in the supporting documentation. The way in which these aspects are handled is clearly set out in standard clauses, forming part of the standard form of contract being used.

Building contracts generally

Fixed price contracts

These are contracts where the price is agreed and fixed before the contract is signed. The agreed price is paid irrespective of the builder's actual costs, subject to a set of contract conditions allowing them to recover additional costs caused by agreed circumstances, e.g. variations or alterations to the contract documents on which his price was based.

These types of contract can be considered to be risk bearing, a degree of risk being apportioned to each party. Fixed price contracts are divided into two types.

(1) Lump sum contracts

With these contracts, the contractor agrees to execute the whole of the work for a stated 'lump sum', which is based on firm quantities, specifications and drawings. There are two derivatives; firm price contracts which do not allow for contract prices to be adjusted for fluctuations in market prices of contractor's costs; and fluctuating price contracts, which provide for the use of a formula for the adjustment of prices to take account of these movements.

Lump sum contracts presuppose that there is sufficient information available, prior to the tender stage, to permit accurate preparation of a firm bill of quantities and specification, and that, therefore, the client's level of risk is minimised through the submission of a firm price for the job. Financial administration of the project is readily facilitated, as a firm basis exists from the outset.

However, in many cases contract documentation is founded on imprecise information and if this happens the presumed advantages may not accrue. In consequence this form of contract, together with traditional competitive tendering methods, has come under increasingly critical scrutiny for new-build work.

Except for major renewal schemes, where firm positive documentation can be prepared, this type of contract is not generally appropriate for maintenance work.

(2) Measure and value contracts

In these forms of contract, the contractor agrees to execute the work, at prices fixed in advance, for units of work to be measured later. These may again be firm or fluctuating, depending on arrangements for coping with market movements in contractors' costs. In some instances the contract documents may include a set of approximate quantities, and this provides the client with a better idea of the extent of his final outlay. However, the indiscriminate use of approximate quantities should be discouraged, as they are often of little more administrative use than a schedule of rates, and may present a misleading picture to the client.

Another variant is the so-called schedule contract, which is used where the work details are too vague to permit the preparation of fully detailed information at the time of its commencement. The schedule, listing all the items of labour and materials that are expected to be used, may be prepared specifically for the job in hand, or be a standard schedule prepared to cover a range of jobs.

The schedule may be unpriced, in which case the contractor is required to enter prices against the schedule items, and this is then used as a means of selecting a contractor. Standard priced schedules are also used, where the contractor tenders by giving a percentage on or off the standard rate. For example, the former Property Services Agency (PSA) of the DoE made great use of a standard pre-priced schedule. The work is measured at completion, or interim stages, by using the relevant rates.

Although a convenient way of providing in advance for the pricing of work of uncertain content, standard priced schedules lack estimating accuracy. Contractors tendering are required to state their offer in percentage on or off the schedule rates as a whole, which assumes that an overall percentage adjustment will bring them into line with those normally charged by the contractor. This is unlikely to be the case for a number of reasons.

❑ There is little uniformity among contractors regarding the pricing of individual items of work, and inevitably some of the schedule rates will be higher and some lower than a given contractor would nor-mally use.
❑ The mix of items for a particular job may well exaggerate the effects of such differences in pricing patterns. For example, a job may consist largely of those items for which the standard rates are higher than the contractor's usual ones and the addition of a percentage would only serve to heighten such discrepancies.
❑ The final cost will depend upon both the quantity and the rate for each item in the work executed. If one accepts that some of the schedule rates are high and some low, then the relative quantities of each influences both the outcome and cost predictions.
❑ The schedule rates are essentially averages, and may not reflect the particular conditions under which the relevant work items have to be carried out.
❑ Standard rates tend to need updating at more frequent intervals than actually occurs in practice.

In order to arrive at a realistic estimate, the tenderer should estimate the likely proportions of schedule items in the job at hand, and then determine a percentage adjustment which will equate the cost based on the schedule of rates to the cost he would obtain if using his normal rates.

However, schedule contracts do fulfil a useful role for many types of work, notably maintenance.

Cost reimbursement contracts

In these types of contract the contractor is reimbursed for the actual prime costs of labour, materials and plant used plus, either a previously agreed percentage, or a fixed fee, to reimburse him for his management costs, overheads, and profit. In its pure form, the disadvantage of this type of contract is the absence of an incentive for the contractor to keep his costs down. Its use is therefore restricted to small or urgent jobs where the necessity to execute work very rapidly provides insufficient time to produce precise documentation, and the degree of risk to the client can be justified.

This type of contract requires a reputable contractor who is known to the client. Indeed, if the client is a good and regular customer, this should help to limit the risk, provided there is a proper checking mechanism at the end of the job.

The low level of documentation places an increased burden on supervision, and requires a very good communication system to be in place during construction. The client's representative, in particular, has a much more demanding job in protecting the client's interests, and a larger measure of interference with construction methods is justified, and should be expected by the contractor.

The competition element is limited, and depends, in substantive terms, on the contractor's percentage add-on. This in itself may not always be the best guide, as a higher percentage add-on may reflect a more efficient attitude to management of the project, which may lead to a smaller final cost to the client. The converse, of course, may also be true.

The best method of selection is probably on the basis of reputation, which may be fine if the client regularly lets this type of contract to a limited number of contractors, but is clearly problematical when the client has no prior experience upon which to draw.

Another factor that must be borne in mind is the checking procedure to ensure that the client obtains what he has paid for, and also the scrutiny of the final account for future reference. This is a notoriously difficult thing to do, especially when it comes to assessing the number of labour hours included. The situation may be helped by requiring the contractor to submit records of resource inputs, at regular stages throughout the execution of the work, to permit ongoing checking and control.

Use has been made of so-called controlled daywork, where all jobs are pre-estimated prior to placing the order. If the contractor's account exceeds the estimate by more than an agreed percentage, the reasons are investigated. This method has the merit of providing the client with a firmer idea of his outlay, although the pre-contract period may be somewhat lengthened.

Latterly, increased use has been made of cost reimbursement contracts for large projects, let on a management contracting basis. This is generally of little relevance to maintenance contracts but there are some developments that are interesting, in particular the use of target cost contracts, where an initial prime cost target is agreed, along with a management fee and an adjustment percentage. Any variation from the target cost then leads to recalculation of the management fee in accordance with the adjustment percentage. The simple example shown below illustrates the principle.

Initial prime cost target	£60000
Management fee	£ 5000
Adjustment percentage	25%

(a) The prime cost at completion = £62000.

The contractor is paid:

Prime cost	£62000
Management fee	£ 5000
Less an adjustment of 25% of excess of £2000	£ (500)
Total payment	£66500

(b) The prime cost on completion = £58 000.

The contractor is paid:

Prime cost	£58000
Management fee	£ 5000
Plus an adjustment of 25% of saving of £2000	£ 500
Total payment	£63500

Thus, in this form of a cost reimbursement project, there is both incentive and penalty for the contractor, and an adjustment process, which shares costs and benefits between contractor and client.

Fixed price maintenance projects

This approach was adopted several years ago, in an attempt to reduce the large amounts of paperwork involved in executing types of contract in which the administrative costs for small projects may conceivably exceed construction costs.

The method involves agreeing a lump sum, based on analysis of maintenance records, with a contractor for undertaking a range of recurring works of a similar nature, to a specified group of buildings, over an agreed period of time. The contractor then agrees to carry out all the work of the specified types for the agreed contract sum. Tendering is through the contractor quoting a percentage on or off the given lump sum. The stipulated time periods are normally a year, and the contractor is paid one-twelfth of the lump sum each month.

In theory, this approach makes large savings in administrative costs, and payments are simply made on a monthly basis. The checking necessary can be through regular site inspections to ascertain that the work is being executed properly.

It has also been argued that there is in-built quality control, as failure of a contractor to execute work properly will manifest itself at an early stage through a call for the work to be done again. This is by no means certain, unless the same contractors are regularly employed for the same tasks on the relevant buildings. However, if the system is set up properly then a rapid response can be expected, leading to good tenant satisfaction.

The major challenge in using this type of contract lies in establishing the correct lump sum each year. Basing the figure on recorded data will only prove an adequate approach if there is a large enough sample on which to produce a statistically sound figure. In addition to this, as with the use of historical costs for budgeting, real maintenance needs may be underestimated.

The operation of this approach requires high levels of goodwill on all sides, and is facilitated if extremely thorough, accurately costed condition data exists.

Term contracts

Under this type of contract, the contractor is given the opportunity to carry out all work for the client, usually the subject of a works order, for a specified period of time, within an agreed geographical area. The work done is then normally priced in one of two ways:

- ❑ Through the medium of an agreed schedule of rates, when it is termed a *measured term contract*
- ❑ On a cost reimbursement basis, when it is normally called a *daywork term contract*

In some cases, for larger items, a lump sum contract may be negotiated.

In essence, the term contract can be viewed as a specific case of other forms of contract described earlier. They have become increasingly used, in the measured term format, for maintenance work, most notably by the now defunct PSA[1]. PSA figures report that 40% of their minor works expenditure in the years from 1982 to 1989 was through measured term contracts, the next most widely used method being lump sum contracts for works between £25 000 and £250 000 in value.

A BMI report[2], based on a limited survey, suggested that the most common basis for such contracts is a priced schedule. The most popular schedules were those developed by the client themselves, followed by the PSA[3] Schedule of Rates and the National Schedule of Rates[4], that accompanies the Joint Contracts Tribunal (JCT) standard form of measured term contract. The report concluded that, whilst terms varied from one to six years, three years is the general norm. For contracts over a year a fluctuation provision is given, either based on published cost indices or annual updating of the rates. The PSA developed a number of standard schedules, and the normal tendering process would make use of the percentage on or off approach described above.

Measured term contracts have been found to be most beneficial where the client has an on-going need for maintenance, or other minor new work. A pre-requisite is that there is a large enough workload to offer sufficient continuity for economic operations. They are also of value in assuring a rapid response to work requests. To ensure this happens, it will be normal to specify response periods, and this requires a good communication system, and means of monitoring and controlling performance.

In theory, the benefits of an assured work programme should lead to keen pricing. However, this may not mean the lowest price for every rate, and hence the lowest price for every order. It must be viewed as a package, and judged in terms of its ability to give the lowest overall price.

Under normal operation, the percentage on or off should allow for fluctuations over the period of the contract. This process is rather uncertain, as wage rates and material prices will change at a different rate, and as their proportions in the workload are not readily predictable, the probability that such an adjustment is accurate is very low. Very large clients may overcome this by regular review of the base rates. For example, the PSA were able to do this through monthly published percentages. Conversely, the reasonableness of contractor's rates in open competition is tested infrequently.

Perhaps the major benefit to be realised, through the letting of an assured programme of work, is the opportunity for the development of a good client/contractor working relationship, with corresponding improvements to quality of service.

It is also claimed that the measured term contract offers potential savings in time and administrative costs, compared with the letting of a multiplicity of contracts. Whilst this may be true for the pre-contract stage, it is a doubtful claim when total costs over the life of the project are considered.

Difficulties may also be encountered at a changeover point, where the contractor for the previous period is being replaced. There will be inevitable disruption to work programmes. The sanction retained by the client, to terminate a contract under given circumstances, may also cause severe disruption, and can leave a large number of incomplete jobs. These factors may place pressure on the client to retain a contractor under circumstances where he should properly be replaced.

Contractor selection

Depending on which contractual form is adopted, there will be a procedure for selecting a contractor. In any type of project there are certain good practice rules that should be adopted, in order to safeguard the client's interests. The objective is to select the contractor, from those available, who is most likely to satisfy the client's requirements. Price will only be one of these requirements, and the specific characteristics of maintenance work give rise to other performance parameters. These largely stem from the need to provide a service, rather than a specific product, and this distinguishes maintenance from other types of work. The ability of the contractor to be able to respond promptly, and to efficiently execute a range of small tasks, often on a frequent basis, but without the benefit of much forward planning, is of major importance.

Up until comparatively recently few contractors specialised in maintenance contract work. However, diminishing new-build workloads have raised the interest of a wider range of companies, who previously would not have considered this type of work. Historically, maintenance contract work was attractive to small contractors, who were considered to be the ones most appropriately structured. The interest of larger organisations in maintenance work is attributable less to the attractiveness of maintenance work than to difficulties in other sectors. Therefore, whilst the increased competition for work may present benefits on a cost basis, there is a need for caution.

The selection of a contractor should be made on as wide a basis as possible, taking care to compare the contractor's known or assumed abilities with those required. The following are general factors that should always be taken into account in contractor selection.

(1) Reputation

Familiarity with a contractor's capability for executing the type of work under consideration is always a valuable starting point. Volatile market conditions impose the need to give careful consideration to this factor, particularly with new contractors seeking to break into the maintenance market. It may well be prudent to question motives, as the contractor may see this market as a temporary one to make use of spare resources. Evidence of real commitment is not easy to establish.

(2) Resources

It is crucial that the contractor possesses sufficient resources to execute the work but, in the case of maintenance work, it is important to consider type and quality, as well as quantity. Much can be learned about a contractor's ability by studying his physical resources, as represented by his stores, offices and workshops, and about his labour resources from previous contracts. Of particular importance is his management set-up, and careful scrutiny of its ability to handle the specific requirements of managing maintenance work is essential.

(3) Workload and availability

Whilst this must always be considered when selecting a contractor, it must be borne in mind that the nature of maintenance work may demand a service measured in terms of response times. This requires a rather different type of commitment from the contractor than on new-build work, where he is able, within certain limits, to plan his labour commitment for a project.

(4) Price

In many cases this is the sole criterion used for judging a contractor, and this view is clearly a very narrow one. It must be remembered, however, that in many cases, particularly in the public sector, accountability is a prime requirement, and that not only is there a need for the placing of contracts to be fair, but to be easily seen to be so. This can often be used as an excuse for an oversimplified approach. Recent developments in housing management by housing associations, where a range of performance criteria are employed, is encouraging in this respect[5].

Selection procedures

A simple way of classifying procedures for selecting a contractor is by the degree of competition.

(1) Open tendering is a method whereby a contract is advertised, and all contractors are free to tender, without any prior enquiry with respect to their ability. This type of tendering is now rarely used, even where accountability is a major consideration, due mainly to quality and performance concerns.

(2) Selective tendering is a refinement of the above method, designed to remove its major drawbacks[6]. Tenders are invited from a list of contractors, which has been compiled having regard to their reputation and, just as importantly, suitability for the type of work in question. Large client organisations may hold permanent lists, which are updated at intervals, but in some cases special lists are compiled for a particular project. In the latter case, an advertisement will invite contractors to apply to go on a list.

(3) Negotiated contracts involve the client inviting a tender from a known contractor, who is judged to have the necessary qualities for the task at hand. This tender may then be subject to further negotiation regarding conditions and prices. Whilst the level of competition, particularly in cost terms, is strictly limited, it can provide a very satisfactory outcome where other requirements, such as time and quality, are considered to be of significant importance.

Tendering can be carried out in a single stage, based on one set of contract documentation, usually at the end of the design stage. So-called traditional competitive tendering for a lump sum contract typically operates under this process. Measured term contracts, similarly, can be single stage.

In some cases, multi-stage tendering may be deemed to be appropriate. For example, in new-build work, the contractor may be selected by means of a first stage competitive tender, based on documentation related to preliminary design information, and which provides a level of pricing for subsequent negotiation. The second stage consists of accurate pricing of the completed design, the contractor having been involved in design development following a first stage appointment[7]. This is a procedure to be followed when it is considered beneficial for the contractor to be involved in the design process.

From the legal viewpoint, the tender represents the offer component of the contract. This offer may be made at any one of a number of points in time, and may be based solely on price or other factors. The offer may be in the form of a fixed price, either stated as a lump sum at the time of tendering, or to be arrived at on the basis of a schedule of rates after completion, or for an indeterminate sum of money arrived at through cost plus a percentage, or fee. The offer may relate to a single job, or to a series of jobs over a defined period of time.

Following the offer, the client or his representatives must make a selection, or adjudication as it is commonly termed. This process may be simple, if price is the sole criterion, but if offers are to be judged on a range of issues then it is important that the relative importance of these criteria is clearly defined. This may be done in an informal way, based solely on judgement and experience, or recourse may be made to a weighting system[8].

It is not essential for there to be a written contract, provided the requirements of common law are met. For example, small jobs are often carried out on the basis of a simple verbal or written exchange. In this case, the absence of written conditions requires the use of some ground rules provided by the following implied terms.

(1)   The building owner must allow the contractor to enter a building at the necessary time for the purpose of executing the work, give necessary instructions within a reasonable time, and not obstruct the contractor in the execution of his work
(2)   The contractor must do the work in a workmanlike manner, and complete within a reasonable period of time
(3)   In addition, there is an implied warranty that any materials supplied are reasonably fit for the purposes for which they will be used, and of good quality

Clearly there is a great deal of uncertainty in the interpretation of these terms and, in the case of dispute, recourse would have to be through common law, with the attendant costs of such actions. For most contracts above a minimum size, the contract conditions should be in writing, and there is a range of standard forms of documentation that can be used for this purpose.

Contract documents

Contract documents are those 'which can be identified as containing the terms of a concluded contractual agreement between the parties which was made in, or reduced to, or recorded in writing'[9]. Documentation for building and construction contracts usually consists of:

❏ Articles of Agreement
❏ Conditions of Contract

and one or more of the following, depending on the type of contract:

❏ Drawings
❏ Specifications
❏ Bills of quantities, either firm or approximate
❏ Schedules of rates

The articles of agreement represent the core of any particular contract, and contain the definitions of the parties, a description of the scope of the works, the price mechanism, and the persons to operate under it. It is here that the signatures of the parties and witnesses go. In most standard forms of building contract the articles can be obtained as a pre-printed form, usually along with the conditions, with spaces left for relevant infill.

The conditions of contract contain the detailed provisions as to how the contract is to be operated. For example, the rights and duties of the parties are more clearly defined, and detailed arrangements given for the treatment of variations to the contract, methods of payment, etc.

A major component of the contract will be price, and the contract conditions must provide proper safeguards for all parties. The concern of the building owner will be that they pay a fair price in relation to the quantity and quality of service provided. Payment mechanisms must, therefore, be carefully laid out to help ensure that this happens, and it is also useful if they assist the operation of a proper cost control system. The various standard contract forms have ways of dealing with this, often making using of pro-formas, perhaps as an appendix to the contract form.

In general there will be conditions relating to each of the following:

(1)   A statement of the lump sum, or the formula which is to be used in establishing the final cost. In the case of a lump sum, there should be an itemised breakdown of the work, with each item priced to assist cost control, and to form the basis for evaluating any variations to the contract. At the limit this is a bill of quantities. Such an itemised breakdown is unlikely to exist for maintenance work.

(2)   A statement of the documentation that must be provided by the contractor to support an application for payment. In a daywork term contract, for example, this will probably include time sheets and material invoices.

(3)   The periods at which the contractor will be entitled to receive payment for work completed. For lump sum contracts this will be

based on an interim valuation of the work. For measured term contracts it may be on a monthly basis, determined by the value of orders executed.

(4) Procedures should be laid down for varying the work, either in terms of quality or scope and, particularly in lump sum contracts, a proper procedure for ascertaining their value. For schedule or daywork based contracts payment should theoretically be automatically received. Problems occur with schedules where an item is not covered, and the contract should make provision for handling this. From a legal point of view, the major problem that occurs is when the scope of the work is altered to such an extent as to bring into question the continued validity of the contract.

(5) In reimbursement contracts, with a fixed or percentage fee, it is necessary to provide a statement as to which costs are prime and which are part of the fee, i.e. deemed to be included in the add-on.

(6) Reimbursement contract forms will commonly differ from other forms, especially lump sum ones, in the scope they allow the client to influence working methods and practices.

Standard forms that are available generally divide into those with, and those without quantities. The JCT have produced a portfolio of contracts, which are widely used for a range of different sized projects. JCT 80[10], which is available with full quantities, approximate quantities, or without quantities, is widely used, primarily for new-build projects of significant size.

The JCT Intermediate Form of contract[11] was introduced with the objective of catering for smaller projects, where JCT 80 was considered to be too complex, although there is some concern that this form is being widely used for projects much larger than that for which it was designed, because of its relative simplicity.

There has been some debate concerning the proliferation of forms of contracts and, in particular, many doubts as to whether they are being used in an appropriate way. Selection of the right one for any project is fraught with difficulty, but there are a number of forms available that are generally appropriate for maintenance work.

JCT Agreement for Minor Building Works

The Agreement for Minor Works is issued by the JCT[12] for use where a lump sum has been agreed, and where an architect/supervising officer (SO) has been appointed. The preamble to this form explicitly states that it is not suitable for use where a bill of quantities has been prepared, for works of a complex nature, or where it is intended to nominate sub-contractors or suppliers. Despite these restrictions, there exist many situations where this form can be used for sizeable projects.

Where the full extent of a maintenance contract can be clearly defined, for example under a planned maintenance programme, then this standard form is quite usable. The main restriction to its use is clearly the complexity of the work, but where it consists of straightforward traditional construction, the provisions provide a good basis for satisfactory administration of the work.

Its use does not preclude the production of a bill of quantities to assist in arriving at a lump sum, but means other than the bill will be required for the operation of the contract. In other words such a bill will not be a contract document, and the contractor should therefore treat it with caution. It is envisaged that a priced specification or schedule will be required to value variations.

Similarly, the contractor is not precluded from sub-contracting elements of the work, nor is the architect or SO prevented from specifying, in the relevant contract documents, that certain specialist operations or materials are to be used. Such sub-contractors and suppliers would then operate on the same basis as the contractor's own.

The time span under which this form can operate is also strictly limited, as it makes no provision for fluctuations.

The form starts with the agreement followed by conditions, and is concluded with a three part Supplementary Memorandum.

The agreement names the parties and identifies the work to be carried out, with reference to the contract documents, which may include one or more of drawings, specification or schedules. The parties to the contract are to sign the contract documents, and the contractor is to price the specification or schedules, or provide a schedule of rates. The contract sum is given in the articles, along with the name of the architect/SO, and there is provision for the naming of a quantity surveyor.

A list of clauses is then given, some of which are asterisked to indicate that they contain blanks to be filled in, or that the clause is optional and may be struck out. The following are examples:

2.1   Commencement and completion, requiring the insertion of the relevant dates.

2.2   Damages for late completion, with the opportunity to insert the amount of liquidated damages.

2.5   The defects liability clause requires the insertion of an appropriate period.

3.6   This important clause deals with variations, and it is required to specify if this is to be through a priced specification, priced schedules, or a schedule of rates.

4.2   The progress payments and retention clause provides for the insertion of the amount of retention, if other than 5%.

Whilst the above clauses represent the fundamental administrative framework for the contract, brief comment on the general purpose of all the clauses is worthwhile.

Clause 1 gives conditions with respect to the intentions of the parties, and includes a requirement for both parties to discharge their duties in a proper way. For example, the contractor has to carry out the work:

'... with due diligence and in a good and workmanlike manner ... in accordance with the contract documents...'

These are the contract documents that have been specified in the agreement and signed by the parties.

Clause 2 is concerned with commencement and completion, and contains provisions for liquidated damages and extensions of time. These are simple in this form. For example, it states that extensions of time may be granted: '... for reasons beyond the control of the Contractor...'. This is a very broad statement, and it has been held by some commentators to be potentially troublesome, and it is wise to restrict its application, by listing what these reasons may be.

It is also clear that there is an onus on the contractor to notify the SO immediately it becomes apparent that the works will not be completed on time. He must then respond by granting a reasonable extension of time, provided the reasons for the delay are valid in this respect.

This clause also contains provisions for defects liability.

Clause 3 is concerned with control of the work, and there are several interesting provisions. If the contractor wishes to sublet any portion of the work, he must obtain the express permission of the SO so to do, although this may not unreasonably be withheld.

Of great importance is the granting of apparently unrestricted power to the SO to issue instructions with which the contractor must comply, and these instructions may be not only in respect of additions and omissions, but also related to the order, or period in which the works are to be carried out. However, it is debatable as to whether this right extends to directing the contractor to execute the work in a particular way. There seems to be a fine dividing line between these two positions.

The power to alter the period of the work, which presumably means in either direction, is also given. This is quite separate from the extension of

time provision and such changes, '...shall be valued by the Architect/
Supervising Officer on a fair and reasonable basis...'.

Whilst there is no provision for PC sums, as nominations are expressly
excluded, there is the facility for provisional sums and these are covered
under 3.7.

Clause 4 deals with the important question of payment. The correction
of inconsistencies is covered in 4.1 and, where these involve a change,
they are treated as a variation.

Provision is made in 4.2 for interim payments to be made to the
contractor monthly, if so requested, less the agreed retention figure. It is
recognised, through 4.3, that final valuation of the work may take some
time, and not be available at the time of preparation of the penultimate
certificate. This clause therefore says that the SO shall:

> '...within 14 days of practical completion, certified under clause 2.4
> hereof, certify payment to the Contractor of 97.5% of the total
> amount to be paid to the Contractor under this contract so far as that
> amount is ascertainable at the date of practical completion...'

However, the final certificate clause places an obligation on the contractor to produce the documentation, reasonably considered necessary
to determine the final amount, and that the final certificate is to be issued
within 28 days of its receipt, or of the issue of the Certificate of Making
Good Defects, whichever is the later.

Although fluctuations clauses are excluded, 4.5 allows for the contractor to be recompensed for changes in levies and taxation. Supplementary Memorandum A gives detailed provisions for this.

Clause 5 covers statutory obligations including, in 5.3, the normal
requirements with respect to tax deduction, which are detailed in Supplementary Memorandum C. Value added tax is also dealt with here,
subject to detailed provisions in Supplementary Memorandum B.

Injury, damage and insurance is dealt with under *Clause 6*, and the
main things to note are the provisions relating to existing buildings. It is
also worth noting that the amount of insurance cover the contractor is
required to take out should be related to the proximity, nature and use of
adjoining premises, and presumably this should include condition. The
size of the job does not necessarily relate to the potential size of claim for
which the parties might be liable.

Clause 7 covers determination, giving the employer the right to dismiss
the contractor if he, '...without reasonable cause fails to proceed dili-
gently with the work...' or if the contractor, '...becomes bankrupt ...
or has a winding up order made...'.

Unlike other forms of contract, the employer is not required to issue a warning notice. The most contentious of the two reasons for determination is the former. The employer must clearly make sure that he is on firm ground, otherwise he lies open to an action from the contractor and subsequent damages. It should be noted that the clauses refer to the determination of employment of the contractor, and the contract remains in force.

The contractor is given the right to determine his own employment if the employer defaults in a number of ways, such as failing to make a progress payment or to make the premises available, or becomes bankrupt, or interferes or obstructs the execution of the works.

JCT Standard Form of Measured Term Contract

JCT have produced a Standard Form of Measured Term Contract[13], which includes a form of tender, as well as articles of agreement, contract clauses and a supplementary section.

The form of tender is described as being for use with the National Schedule of Rates, although the remainder of the standard form permits its use with other schedules. There are other measured work forms used by large clients, which tend, however, to use the JCT form as a model.

The form of tender has provision for the entry of the following information:

❏ The employer's name
❏ The contract area
❏ The name and address of the tenderer
❏ An offer section
❏ A statement that indicates the length of time for which the offer will stand

The offer has several constituent parts. The tenderer offers to carry out, and complete, all orders for work, in accordance with the information set out in the contract particulars, and with the conditions of the JCT Standard Form of Measured Term Contract and the rates listed in the National Schedule of Rates. The latter is subject to the addition/deduction of what is termed the *percentage A*, which the contractor enters in the appropriate space. This represents the contractor's basic offer.

The contractor is then asked to insert *percentage B*, which is the percentage he requires to be added to works carried out as daywork under the contract. A percentage is to be entered against each of the following items under this heading:

- ❏ Overheads and profit on labour
- ❏ Overheads and profit on materials
- ❏ Overheads and materials on plant
- ❏ Overheads and profit on sub-contractors

He is also required to enter, under this section, the basis to be used for varying the labour rate.

A *percentage C* entry is then required for overheads and profit on non-productive overtime.

Articles of agreement

The articles of agreement of the contract comprise the names and addresses of the parties, along with their signatures. There are then three articles, the first of which requires the contractor to carry out maintenance and minor work, set out, or referred, to in the appendix, subject to the conditions of the contract. In the second, the employer undertakes to pay the contractor in accordance with the conditions of the contract. The third article defines who the contract administrator (CA) referred to in the contract conditions shall be.

Appendix A to the standard form

Appendix A to the standard form is divided into several sections, and sets out a number of essential parameters for the administration of the contract.

Section one requires the entering of a list of the properties, in the so-called contract area, in respect of which orders can be issued under the contract. It also requires a description of the type of work for which orders may be issued.

The contractor's right of access to the site, which may be an individual property, is covered under contract clauses 3.4 and 3.5. In essence, the contract conditions state that it is the responsibility of the CA to arrange access, unless it is stipulated otherwise in the preliminaries to the schedule of rates. There has been some criticism of this condition, on the grounds that it is normally more realistic for the contractor to arrange his own access, and that this should be the norm contractually, i.e. the standard form has it the wrong way round.

Section two specifies the maximum and minimum values of orders that may be issued under the contract. In some variations on this standard form, the tender invitation provides value bands, and gives the contractor the option of putting a different percentage on or off (*percentage*

A) against each band. It has been found in practice, however, that given this option, contractors still have a tendency to use a single percentage.

Section three gives the approximate anticipated value of the work to be covered by the contract. It is important to note, however, that under clause 1.13 the employer accepts no responsibility as to the actual amount of work that will be ordered, and that no change in the *percentages A, B and C* will be considered by the Employer if the actual value of the work ordered differs.

Section four specifies the contract period and its commencement date. It is suggested that this should never be less than a year, nor longer than three years. Clause 1.15 requires that the contractor must provide the contract administrator with a programme for the works, if so requested.

Under clause 2 of the conditions, it is a requirement that orders shall be reasonably capable of being carried out within the contract period, unless otherwise agreed. It also asserts that orders must be executed in accordance with any priority coding specified. The details of such a priority code must also be given in section 4 of the appendix. Each order that is given will require completion by a certain date, and there must be a mechanism for agreeing that an order is complete. The contractor notifies the CA when it is considered that this is the case, and this date stands unless the latter disagrees within 14 days.

Clause 2 also provides for delays, and requires the contractor to notify the CA, when it becomes apparent, of any matter likely to cause completion of an order to exceed the specified completion date. The clause refers to matters beyond the contractor's control, and provided the CA is satisfied, the contractor may then fix a new completion date accordingly. Note that, if this goes beyond the contract completion date, the contractor must still discharge his duties under the contract. This is referred to as 'fixing later date for completion', and is not an extension of time to the contract.

The schedule of rates to be used for the contract is referred to in *section five*, but actually specified in *section six*. *Section six* allows for the entry of *percentage A* referred to above.

Sections seven and eight represent alternatives, depending on whether or not a provision is to be made for fluctuations. In the event of no fluctuations *section seven* is deleted. If fluctuations are to be provided for then *section eight* is deleted and *section seven* specifies the basis on which rates may be varied. In the case of the National Schedule of Rates being used, then they are updated on 1 August, and any orders issued after that date may be valued using the new rates. If a different schedule of rates is used, then an appropriate formula will need to be entered.

Section nine provides for entry of the *percentages B and C*, in respect of daywork and overtime working, and the appropriate labour rate. Clauses 4.4 and 4.5 lay down the procedures for the measurement of daywork[14]. This states that:

> '...it shall be calculated in accordance with the definition of Prime Cost of Building Works of a jobbing or maintenance nature, subject to adjustment by percentage "B".'

The contractor is required to give the CA reasonable notice of the commencement of any work, or material supply, which he considers should be paid for on a daywork basis, and must then submit returns, in a form required by the CA, within seven days of the end of the week to which the work relates. The standard form specifically notes several instances where daywork payment is appropriate.

❑ Under clause 4.3, where it is not practicable, or would not be fair and reasonable, to value work using the schedule of rates, or to deduce appropriate rates, then payment on a daywork basis is justified.
❑ Where the contractor's access to execute the work is disrupted, then he may, under clause 3.4.2, recover the cost of unproductive time on a daywork basis. Similarly clause 4.7 deals generally with disruption to the work, and payment by daywork is specifically referred to.
❑ When an order has been cancelled, then the contractor may also recover abortive costs through daywork, under clause 3.8.2.

Additional payment for non-productive overtime, and the addition to labour rates of *percentage C* is permitted under clause 4.6, which lays down the rules for this type of payment, and specifies the contractor's responsibilities in this respect.

Clauses 4.8 and 4.9 of the standard form define the responsibility for measurement of the work. In the interests of simplifying administrative procedures, it states that work under a certain value can be measured by the contractor. Above this value, the CA is responsible for valuation of the work.

The appropriate value at which this occurs is entered in *section ten* of the appendix. It is noted in 4.10 that this may be deleted, in which case the contractor effectively becomes responsible for measurement of all the work. The detailed rules with respect to payment and dissent for either of these scenarios are covered by 4.13 and 4.14.

Clause 4.11.1 allows for progress payments to be made on orders over a certain value. This amount is also entered in *section ten*. For orders that fall below this level, then 4.12 provides for payment to the contractor within 14 days of them notifying the CA that the order is complete. This is subject to the more detailed conditions on responsibility for measurement of the work.

Variation or modification of the design, quality, or quantity of the work, or supply, comprising an order, is subject to the rules laid down in clause 3.6.

Under 3.8, the contract administrator may cancel an order, and this clause also provides for reimbursement of the contractor for the costs he may thus have incurred.

Sections eleven, twelve and thirteen deal with contractor's safety policy, statutory tax deduction, and insurance respectively. The latter requires the entry of an amount of cover, percentage addition for professional fees, and the annual renewal date.

Clause 8.1 states that, notwithstanding the duration of the contract period, the employment of the contractor may be determined, by either party, not earlier than six months from the commencement of the contract period, provided that a notice of 13 weeks, or some other period specified in *section fourteen* of the appendix is given.

This so-called break clause is considered to be very important for this type of contract. The smooth operation of measured term contracts requires that the parties develop a good working arrangement. It was felt, when the standard form was being drafted, that this 'escape' clause was necessary to allow dissolution of the agreement, if it became apparent that a good working relationship did not exist.

Section fifteen permits the entry of a named arbitrator to deal with disputes arising under the contract.

Other important features

There are several other aspects of the standard form that are worthy of further comment.

(1) The contract conditions do not contain any preliminaries or specification. These are normally incorporated into the schedule of rates, although they may form a separate document.

(2) The contract form does not allow for nominated sub-contractors or suppliers. Clause 3.2 requires the contractor to obtain the express approval of the CA, prior to sub-contracting any work. In normal parlance, '...consent shall not be unreasonably withheld or delayed'.

(3) The contract gives the right for the employer to supply materials or plant for any work under 1.6, and the subsequent clauses lay down the ground rules for the ownership and responsibility for storage,

and general well being of any such item. This permits large organisations to keep a stock of standard items, and presumably take
advantage of bulk purchasing of heavily used items. This may be on
economic grounds, perhaps to enhance public accountability, or to
guarantee the availability at all times of critical components.

(4) Clause 1.2.2 gives the employer the specific right to place orders for
similar work within the contract area, and within the contract time,
with any other organisation, including his own labour. This is
probably a sensible safeguard in the eventuality of a contracted
organisation being over committed, perhaps due to exceptional
circumstances.

(5) All orders have to be in writing, which may include drawings. No
oral instructions can be given, although there is the provision for
orders to be confirmed in writing following a verbal order.

(6) There are no specific provisions for liquidated damages and, at the
time of writing, it has yet to be tested as to whether an employer
may recover losses due to late completion at common law.

(7) There is a six months' defects liability period, but no retention
money under the standard form. It does however permit the
employer to recover the cost of carrying out remedial work, not
done by the contractor, from future payment.

The question of insurances is sufficiently important to merit special
attention, and the contract provisions under this standard form provide
a good example of the general requirements in this respect. The matter is
dealt with under *section six* of the standard form, and can be considered
under several headings.

(1) Injury to persons and property and indemnity to employer
The contractor is required to indemnify the employer against liability,
etc., arising under statute or common law, as a result of personal injury
or death, caused by the carrying out of the order, except where this is
specifically attributable to the actions of the employer, or persons for
whom he is responsible. Similarly the contractor has to indemnify the
employer of losses to persons or property arising due to negligence,
breach of statutory duty, omission or default of any of his employees, or
any other person on the site for the purpose of executing the order, with
the exception of the employer or his representatives. The reference to
property does not include the work comprised in the order.

These provisions summarise clauses 6.1, 6.2 and 6.3. Clause 6.4 goes
on to say that, without prejudice to these responsibilities, the contractor
must take out an insurance policy to comply with the Employer's

Liability (Compulsory Insurance) Act 1969. It is required that the insurance cover must not be less than the amount inserted in *section thirteen* of the appendix.

The contractor is required, under the contract, to provide documentary evidence that insurance has been obtained. If this is not done, then the employer may take out the relevant policies, and recover the cost of doing so from the contractor.

The effect of ionising radiation, or contamination by radioactivity from any nuclear fuel or waste, is specifically excluded under 6.4.5.

(2) Insurance of existing structures

This is clearly of major importance in maintenance work, and under this standard form, clause 6.5 requires that the employer should insure the existing structures in respect of which orders are made, together with the contents which they own, or for which they are responsible, against loss or damage due to fire, lightning, explosion, storm, tempest, flood, bursting or overflowing of water tanks, apparatus or pipes, earthquake, aircraft and other aerial devices or articles dropped from them, riot and civil commotion. The exclusions under 6.4.5 are also relevant under this heading.

Provision exists for joint insurance by employer and contractor, and rules are given for the apportionment of costs.

Except when a local authority, the employer is required to provide the contractor with documentary evidence of the relevant policies as requested.

(3) All risks insurance

The contractor will normally be required to take out an all risks insurance policy, which provides cover against any physical loss or damage to work executed, or materials supplied, with respect to any order made, and any site materials. Under the JCT Standard Measured Term contract, the latter may include any materials supplied by the employer.

Under 6.8 in the standard form this specifically excludes:

- Property which is defective due to:
  - wear and tear
  - obsolescence
  - deterioration, rust or mildew
- Loss or damage due to defects in design, specification, material, or workmanship
- Loss or damage due to war, invasion etc., radiation etc., as outlined

in 6.4.5, and any unlawful act executed by an unlawful association, i.e. terrorism

The clauses 6.9 to 6.12 give detailed requirements for all risk policies, and the ways in which claims should be handled for a variety of circumstances.

A Standard Local Authority Form of Measured Term Contract

A number of large estate owners have their own standard forms. One example is that used by Leicester City Council, called a 'Schedule of Rates Contract for Building Works'. This form is clearly designed for more than just straightforward maintenance work and is, in many respects, more extensive in terms of its conditions than the JCT standard form, although the principles are similar.

The form used divides into the following sections:

❑ Invitation to tender
❑ Conditions of contract
❑ Abstract of particulars

The abstract of particulars is equivalent to the appendix of the JCT form and its contents are as follows:

❑ Facilities available to the contractor
❑ Appendix A – domestic specialist sub-contractors
❑ Appendix B – list of buildings/properties
❑ Alphabetical index to conditions of contract
❑ Form/s of tender
❑ Tender addendum
❑ Map of contract area

Under the invitation to tender, the council specifically indicate that they will either nominate, or enter into separate contracts for electrical work, heating, ventilation, and domestic engineering work. The contract conditions, therefore, have provisions to deal with both nominated and domestic sub-contractors, and the resolution of problems at the interface of contractor and directly employed personnel.

The invitation to tender also indicates that the employer reserves the right to enter into separate contracts for large painting or redecoration works, roadworks, landscaping, or other such sundry works which it considers as not being appropriate to, or a part of an order for building works. This is clearly a mechanism to permit the council to maximise its options with respect to utilisation of its DLO.

The definition section outlines the meaning of the most important terms. There are minor differences here. For example they refer to and define 'The Supervising Officer' or SO.

Under *section 5* they define the scope of the works, and specifically list the following, thus clearly indicating the larger range of applicability of this form:

❑   Building and civil engineering maintenance work
❑   Alterations and extensions to existing buildings
❑   Minor new works of a building or civil engineering nature

The abstract of particulars, attached to the contract, specifies the minimum size of any order that can be made. There is no provision for measurement and valuation of work of any size by the contractor alone. Provision is made for interim payments on orders over the sum of £5000, which is specified within the contract conditions. The contract also provides for a retention by the employer of 5%.

The abstract of particulars specifies a minimum and maximum contract period, with the employer having the right to terminate the contract at any time after the expiry of the minimum period, provided a notice of eight weeks is given.

There are more explicit conditions laid on the contractor with respect to the personnel and labour employed on the contract, for the use of the site, materials, works and workmanship, removal of rubbish, watching, lighting and protection of the works, existing services, drying out the building and the facilities (including plant) and accommodation to be provided.

Other clauses spell out very clearly the contractor's obligations under safety and welfare, and significantly, *section 35* details a series of requirements under noise control.

There are special clauses dealing with unloading and loading, electricity and water for the works; mixing, deposition, and storage of materials on the highways, and reinstatement of them. The Industrial Training Act is also specifically covered by a contract condition.

Many of the conditions in this contract form are analogous to a preamble section for a normal contract. There are other additions which reflect its potentially wider use, and some which are a reflection of what the authority considers its responsibilities to be in a wider sense.

Jobbing Agreement

This is a comparatively new form of contract, introduced by JCT in April 1990[15]. It has a set of very simple contract conditions, covering about three sides of A4. It is capable of being used with the JCT standard form of tender, or with an organisation's own form. For example, a local authority or housing association's standard works order is likely to be more appropriate than the rather lengthy JCT standard form of tender invitation.

Use of the JCT tender invitation requires it to be accompanied by a full description of the work required, with a specification and/or drawings. For many of the jobs, for which this form of contract is intended, this approach may be inappropriate. The following information is required to be entered on the tender form:

- The start and finish date
- The minimum amount of public liability insurance
- The defects liability period
- The name of the employer's representative

It is also possible to place a verbal order by telephone, provided this is followed by a written order, the major stipulation being that fair and reasonable time periods are allowed, and that the price is reasonable. This scenario presupposes that the employer has a list of approved contractors, who are already in possession of a copy of the contract, and have previously agreed fundamental issues, such as the work area and insurance coverage.

Contract documentation

The object of contract conditions should be to provide a fair and equitable legal framework, which will ensure that the work is carried out in a proper manner, and that the contractor will receive reasonable payment. It is not intended to deal in detail with the legal relationships between the parties, but merely to tackle those features which have a particular relevance in the context of the work being executed.

It is, however, an essential requirement that the parties should have a clear understanding of their responsibilities and obligations, and there is, therefore, a need for supplementary documentation, which may take a number of forms, depending on the type of contract considered to be most appropriate.

The aims of this documentation can reasonably be summarised as follows.

(1) To provide a basis for the submission of a tender, and later evaluation of the work for payment purposes.

(2) To provide a clear picture of the job content or conditions, in a format useful for project management.

(3) The third objective, which is rather all embracing, is to provide a clearly defined supplement to the ground rules laid down by the contract conditions, for the execution of the project.

Contract documentation has to be clear and unambiguous, within the contextual limits that exist, and, most importantly, be in a form that is appropriate for the nature of the work, and the conditions under which it is to be undertaken.

For much new-build work, it is assumed that the full scope of the works is defined through drawings, specification and bills of quantities, prepared prior to obtaining tenders, leading to a lump sum contract.

For maintenance work, it is rarely the case that sufficient information is available to describe the full scope of the work in precise detail prior to tender. Documentation methods may, therefore, vary from a detailed schedule of items to be priced, to a broad statement of the service to be provided, accompanied by a statement of the end result to be achieved. Schedules of rates are therefore of great importance for contract maintenance work, and maintenance specifications will have particular characteristics.

Schedules of rates

The schedule of rates is designed to have similar primary functions to that of a bill of quantities. In the first instance, it provides a means of comparing the tenders from a series of contractors on the basis of price. To fulfil this function easily, some simplicity is desirable. A comparison is easy, for example, if contractors are required to add a single percentage only. If, however, banded order values are given, with the contractor able to put different percentages against each, then comparison is only easy if the relative proportion of work in each band can be estimated easily. In other variants of schedules, items may be grouped in some way, with the possibility of a different percentage for each group. Again, this may lead to problems of comparison when the tenders are being assessed.

In the case of an unpriced schedule, where the contractor enters his own rates, there will be great uncertainty in assessing tenders, unless the relative quantities of every item are known within reasonable limits. For these reasons, schedules for maintenance work are generally of the pre-priced variety.

Schedules may be produced on an *ad hoc* basis, following a similar pattern to a bill of quantities, for a particular job or, as is more usual, produced in a standard form. Schedules are normally prepared to cover a range of repetitive jobs, and may be devised to meet the particular needs of an organisation, or may be standard ones which are more broadly based for general application.

The second function of a schedule is to provide a basis for payment, in conjunction with the contract conditions, and is thus fundamental to the use of the measured term contract.

The decision as to whether to use an *ad hoc* or a standard schedule depends on a great many factors. Standard schedules will generally have the following advantages:

❑ They provide a relatively cheap control document for larger size contracts
❑ Because they are readily available, administrative costs in preparing contract documentation are reduced
❑ They can be an integral part of an established set of procedures, which have been well tried and tested, and fit in well with computerised systems
❑ Both employer's representatives and contractors will be familiar with their use, which reduces the perceived risk of all the parties, and promotes good working relationships

On the other hand they do have some disadvantages:

❑ Their comprehensiveness can make them unwieldy, particularly for smaller contracts
❑ The standard form may not be appropriate to the needs of the particular estate
❑ They need to be able to integrate properly with established management systems

Bespoke schedules may be more relevant in some cases, as they have the advantage of conciseness, and can be tailored to suit the specific needs of a contract, or of a particular client. On the other hand, there may be a lack of familiarity with their use, and this may cause operating difficulties and increase costs. They may also lead to increased risk levels, and lead to a cautious pricing policy by contractors.

The PSA of the DoE produced a series of standard schedules[16] for their own extensive range of work, and these were used mainly with term contracts. These types of schedules provide a basis for estimating costs, tendering and valuation during execution. The PSA schedules were used for new-build, as well as maintenance and repair, and published monthly percentage update indices assisted in the administration of a fluctuation clause.

A National Schedule of Rates has also now been produced jointly by the Building Employers' Confederation and the Society of Chief Quantity Surveyors in Local Government[17]. This is a comprehensive schedule, designed with a bias towards housing. It covers all building trades, and new work, as well as repairs and alterations. The format is much like a bill of quantities, but includes separate material, labour and plant costs. There is also a separate Painting and Decorating Schedule, and a document giving labour constants used in the build up of rates. The National Schedule of Rates is re-issued in updated form each year.

A properly maintained schedule of rates provides other important subsidiary functions.

(1)   The priced schedule enables maintenance managers to estimate, with some accuracy, the cost of a maintenance programme, which assists in the evaluation of alternative strategies and the framing of bids for funding.
(2)   A properly maintained schedule will enable the manager to value each order, and thus provide data for an efficient cost control system. This will further enable them to control the flow of work, and perhaps evaluate alternative priorities on an on-going basis.
(3)   Some types of schedule can be adopted to form the basis of an ordering system, and may considerably simplify administration. The manager may effectively be purchasing a maintenance service, rather like purchasing an item of goods. Theoretically, the order value and invoice can be the same.

The BMCIS originally produced a price book, intended as an aid to estimating maintenance work. This price book has also been used as a basis for measured term contracts. Its use in this way is extremely limited, but it does have the merit of simplicity, under some circumstances.

The contents of a schedule of rates will vary in detail, but essentially should include the following sections.

(1)   A set of preliminaries, which define the scope of the work and fully describe the conditions under which it has to be executed. This will include reference to contract conditions. For example, it may refer to the standard form to be used and any exclusions or modifications that will be made to it. In essence, like the preliminaries section to a bill of quantities, it should bring to the contractor's attention the major issues that will affect the pricing of the work. A set of

standard preliminary clauses will normally be used, but these must reflect the conditions for that specific contract.

If a standard form of contract is being used, such as the JCT Measured Term, then this and its appendices will cover the major items, so that the preliminaries need only be a simple list.

(2) It is necessary to include sufficient specification information to enable materials and workmanship to be properly defined and controlled, although these should be as simple as possible. The larger organisations will carry a set of standard specifications, and there is also a National Building Specification for maintenance work.

(3) The actual schedule of rates can take many forms. The essential requirement of all of them, however, is that they should be relevant to the work in hand.

The normally accepted practice is for the schedule to contain a full description of the item, including labour and materials, with a price attached to it. In the past schedules have been produced where the price is only for labour, with material costs being reimbursed at cost. Schedules of this type will not effectively perform the important subsidiary management functions, unless separate material costs can be easily added in. Clearly, this makes cost control and budgeting more tedious.

It has also been suggested, notably by the Audit Commission[18], that schedules should include some information on the predicted relative occurrence of each item. Whilst this is, in theory, useful for the contractor, it would seem that the benefit can only be fully realised if the contractor is given the option of using a variety of percentage add-ons. This then destroys the simplicity of approach that is generally endorsed by the majority of building maintenance professionals.

An ideal schedule would contain the exact description of the work to be carried out on any order that might be issued under the contract. The range of combinations possible, however, is so large that this is not feasible. In practice a reasonable balance has to be sought, and this requires the use of schedules that are relevant in content, without containing too great a proportion of superfluous items. An item that is rarely used is, on balance, better left out, as the inclusion of unnecessary items can be misleading to the contractor, as well as leading to a cumbersome document.

On the other hand, if relevant items are excluded, then a greater proportion of work will need to be valued by some other mechanism, including daywork. It will also reduce the efficiency of the schedule as a cost management tool.

A balance has to be struck between conciseness and completeness, and there is no panacea for this dilemma. The most important prerequisite to effective schedule usage is a good knowledge of the estate that is being managed and past experience. Two types of schedule structure have been proposed to attempt to tackle this problem.

(1) The use of a very detailed schedule, which contains individual items of work for separate tasks, so that a job can be priced by building rates up from individual items of work, rather like a conventional estimating price book.

(2) A composite schedule containing items representing a range of possible descriptions for an item of work.

If, for example, it is required to replace a length of guttering, the detailed schedule would include a series of tasks, from which the price for replacement of guttering could be derived. On the other hand, a composite schedule would contain one or more items for replacing a length of guttering, including associated ancillary work. More than one composite description may be included for this, to reflect the different conditions under which it may have to be carried out; for example its height above ground level.

The former has the benefit of accuracy and flexibility, whilst the latter has the merit of simplicity, both in terms of pricing and valuation. If the maintenance work is to be carried out on an estate which is fairly homogeneous, in terms of building type and usage, then one would tend to the use of composite schedules, as there is less uncertainty attached to operations. If these conditions are not met then the use of detailed schedules may be preferred, in order to avoid excessive use of daywork, but at the expense of increased administration costs. The size of orders will also be influential. A tendency to large orders provides a better justification for detailed schedules.

Specifications

The presentation of information in a specification can take a wide variety of forms, depending on its use, and whether or not it is to be a contract document. For large new works, the architect may prepare a specification, as part of the original brief, to the quantity surveyor to assist in the preparation of the bill of quantities. For smaller new works, such as extensions and alterations, in the absence of a bill of quantities, the contractor may be required to base the tender on drawings and a specification. For a large proportion of maintenance work there will be no drawings, so that the specification takes on a great deal of significance. It is likely also that it is produced by a client's representatives, rather than by an architect or quantity surveyor.

For maintenance work there are a number of other particular difficulties that should be noted, which will affect the form and content of the specification.

For both planned and preventive maintenance, it can be assumed that some sort of scheme has been prepared, and that the work can be reasonably well specified, thus permitting the use of standard specification clauses. Even on planned work, though, the uncertainty element associated with the nature of repairs may invalidate specification items.

For response maintenance, and repair work which is not predictive, a range of degrees of urgency may be associated with the work, involving the use of a priority coding system. The requirements in this respect may have to be included as part of the specification.

In general the following points need to be considered, when designing specifications for maintenance work:

(1)   Much maintenance work may be of the same nature as that carried out for new-build work, in which case standard specification clauses from a suitable source may be used.

(2)   There will be a great proportion of the work, under a planned or preventive programme, where the extent of the work is clearly defined, and for which standard maintenance specification clauses exist, either in-house or nationally. However, the execution of an item may bring to light the need for additional repair work which is a pre-requisite for the completion of the planned item. Many specialised maintenance clauses are written in such a way as to provide sufficient flexibility to allow for this work to be carried out at the same time. For example, a standard clause for the inspection of rainwater goods may also include an instruction for carrying out repair work. Similarly, internal decoration specifications will normally allow for repairing defective plasterwork, identified as part of the operation.

(3)   There are also operations that fall broadly under the heading of repair work, which in themselves are repetitive, and will, therefore, be the subject of a set of standard specification clauses. Examples of this might include underpinning, injected damp-proofing, etc.

(4)   Much maintenance work is of an unknown quantity or type. Defects often manifest themselves in a number of ways, without the real cause being apparent. Specification clauses will therefore need to consider the work necessary to:

- determine the cause of the defect;
- eradicate the effects of a defect;
- rectify the defect.

The timescale for each of these operations may be different in terms of urgency. Serious water penetration, due to a flat roof failure, may require emergency action to clear up initial effects and to effect a temporary repair, followed by investigation and action to provide a permanent solution.

Whatever its content, it is essential that the specification is clear and unambiguous, and a great deal of care has to be taken to ensure that items which affect the execution of the work are properly and accurately described. For example, where soil conditions are important, then the specification must properly consider them. There are clear legal implications resulting from the production of an inaccurate document.

Assistance in drafting individual clauses is available from the National Building Specification[19], or from the publication 'Specification'. The PSA also produced a Specification for Minor Works[20], and the Building Projects Information Committee[21] have produced useful supplementary advice.

In terms of structure, a specification document is normally divided into three main parts:

(1) Preliminaries

It is often an accepted practice to include the preliminaries as part of the schedule of rates. However, wherever they are included in the contract documentation, the preliminaries give a set of provisions which govern the general conduct of the contract, and the extent of the contractor's liabilities and obligations. These must be prepared very carefully, as they have to meet the specific requirements of the contract in hand. Generally, this section will include information on each of the following matters:

- A general description of the work
- The form of contract to be used, together with details of any modifications to it
- The contractor's obligations with respect to the provision of scaffolding, plant, etc.
- The requirements for the provision of huts, both for storage and for the administration of the contract, and to meet the welfare requirements of the site personnel
- An office for the employer's representative, if there is need for one
- The need to provide water, lighting, and power for the execution of the work

❑ Protective measures to be taken, especially where the work has to take place in occupied buildings

❑ Information regarding times of access to the buildings, restrictions on working methods, and other items that will affect the method and progress of the work

(2) Materials and workmanship

This section describes the quality of the materials to be used, methods of construction and standards of workmanship. Specifications for new works normally follow the order in which the work sections are given in the Standard Method of Measurement of Building Works, or the CCPI Common Arrangement.

For many smaller projects, not all of these sections will be required, and there may also be merit in grouping some sections together. In many cases, it becomes rapidly apparent that a different approach is required. In particular, for alteration and repair work, it is common practice for the ordering of the specification to follow the sequence in which the work is to be executed on site.

Irrespective of the sequence in which the specification is written, the clauses within each section are normally grouped as follows:

(1)   Clauses of general applicability to the work section, which may include general job descriptions, and items such as plant, cleanliness, and material storage

(2)   Materials, and their preparation, and which may be described in the following ways:

❑ A full description, stating desirable and undesirable features, and any tests with which they should comply

❑ Stating the relevant British Standard, which is considered to be a minimum quality, together with any Agrément Certification, and noting that in many cases more than one quality standard may be called for within a material, depending on its application

❑ Specifying a proprietary brand, or naming a particular manufacturer or source of supply

❑ By giving a prime cost sum and an outline description of the material

General adjectives such as 'best' or 'first class' should be avoided, unless they are recognised terms used to describe the particular quality being specified. The term 'other equal and approved' also needs to be treated with some caution.

There is a strong case for grouping materials in a separate section to avoid repeating the description wherever it is encountered, or to eliminate complex cross-referencing. The workmanship requirements would then follow the materials groups.

The workmanship clauses should state precisely the standards expected, giving details of any constraints on the method of working. Reference can also be made to relevant standards and codes of practice. These clauses will normally be produced on an operational basis, i.e. follow the sequence of site operations.

(3) The works

This section describes the works to be carried out, using the materials and workmanship set out in the previous sections. Whereas the information for writing the specification for new work comes from design drawings and schedules, much of the information concerning work for existing buildings must be gathered on site.

A draft specification is best written on site, and this enables each item to be fully considered and followed through. Taking rough notes, and writing descriptions later, takes longer and increases the possibilities of mistakes and errors. This underlines the importance of well structured survey information.

It is helpful if the order of clauses in the work section is presented in such a way that each item is successively met in an ordered walk through the building, and there is consistency with survey methodology.

References

(1) Building Maintenance Information Ltd (1993) *Measured Term Contracts*, Special Report 193. RICS.
(2) Building Maintenance Information (1972) *Measured Term Contracts: A survey of current practice*, Special Report 215. RICS.
(3) Property Services Agency (1990) *PSA Schedule of Rates for Building Works*. PSA (DoE), London.
(4) Society of Chief Quantity Surveyors in Local Government *National Schedule of Rates for Building Works*. Building Employers Confederation, London.
(5) National Federation of Housing Associations (1987) *Standards for Housing Management*. National Federation of Housing Associations, London.
(6) Joint Contracts Tribunal (1994) *Code of Procedure for Single Stage Selective Tendering*. JCT, London.
(7) Joint Contracts Tribunal (1994) *Code of Procedure for Two Stage Selective Tendering*. JCT, London.
(8) Janssens, D.E.L. (1991) *Design-Build Explained*. Macmillan, London.
(9) Powell-Smith, V. & Chappell, D. (1983) *Building Contract Dictionary*. Architectural Press, London.

(10)   Joint Contracts Tribunal (1980) *JCT80 Standard Form of Building Contract*. JCT, London.
(11)   Joint Contracts Tribunal (1992) *Intermediate Form of Building Contract*. JCT, London.
(12)   Joint Contracts Tribunal (1994) *JCT Agreement for Minor Building Works*. JCT, London.
(13)   Joint Contracts Tribunal (1992) *JCT Measured Term Contract* (5th edition). JCT, London.
(14)   Royal Institution of Chartered Surveyors (1975) *Definition of Prime Cost of Daywork carried out under a Building Contract*. RICS, London.
(15)   Joint Contracts Tribunal (1990) *Jobbing Agreement*. JCT, London.
(16)   Property Services Agency (1990) *PSA Minor Works Specification*. PSA, London.
(17)   Society of Chief Quantity Surveyors in Local Government, *National Schedule of Rates for Building Works*. Building Employers Confederation, London.
(18)   Audit Commission for Local Authorities in England and Wales (1986) *Improving Council House Maintenance*. HMSO, London.
(19)   NBS *National Building Specification*. NBS Services, London.
(20)   Property Services Agency (1990) *PSA Minor Works Specification*. PSA, London.
(21)   Building Project Information Committee (1987) *Project Specification – A code of Practice for Building Works*. BPIC, London.

Chapter 8

The Execution of Building Maintenance

Introduction

Of all the considerations relating to the execution of maintenance work, perhaps the strategic issue of who actually does the work is the most important. Traditionally, there have been significant differences of practice between public and private sectors in this respect, although this is changing somewhat. It remains true, however, that many of the principles of maintenance management practice were established in the public sector.

Within this sector, there was a common presumption that maintenance work would largely be executed by a direct labour organisation (DLO), whilst in the private sector, only work of a small jobbing nature was normally carried out by a limited in-house work force. Traditionally, the degree of interest shown in contract maintenance work by private sector contractors tended to be somewhat limited.

The position at the time of writing is more fluid, and this is due to two major factors:

❑ Increasing constraints placed upon the operation of DLOs
❑ Market conditions, in the construction industry as a whole, arousing increasing interest amongst contractors in a broader range of work

The debate with respect to who executes maintenance work in the public sector has been an acrimonious one, with a tendency for the battle lines to be drawn along party political boundaries. There are, however, important managerial and technical issues to be considered that suggest the need for a pragmatic approach.

Direct labour organisations

The growth of the DLO in the immediate years preceding the legislative framework set up in 1980 to govern their activities, was relatively consistent with the growth in local authority activity in general, and the real expansion pre-dates this by some ten years to 1970.

The building boom of that time made it increasingly difficult for local authorities to obtain realistic prices for their building work. This prompted them to set up their own works organisations. The recession that followed the boom increased competition levels, and the only way that the local authorities could continue to maintain a DLO was to give them preferential treatment.

This fuelled the opposition, which argued that there was little justification for them in a healthy, competitive market. The counter arguments are based around the cyclical nature of construction work, and were used as a justification for maintaining the existence of DLOs in the public interest. Increasing pressure on their activities, however, eventually led to the Local Government, Planning and Land Act 1980, the provisions of which presented a two-pronged attack.

(1)   Accountancy reforms, designed to make all DLOs financially autonomous, were implemented so that in consequence their efficiency could be monitored and controlled against yardsticks. To do this, the Act defined four areas of work for which a local authority was required to keep separate revenue accounts. Broadly these are:

   ❑ General highway and sewer maintenance work
   ❑ New construction, other than that in the above, costing over £50 000
   ❑ New construction costing less than £50 000
   ❑ Maintenance other than that in connection with highways and sewers

(2)   The Act conferred reserve powers on the Secretary of State allowing him, at his discretion, to shut down or curtail the activities of any DLO failing to meet a target rate of return on capital employed. Specifically, at the time of the Act this rate of return was set at 5%.

The Act, and later regulations, required that all maintenance contracts over £10 000 be put out to tender under open competition, before being awarded to the DLO. In 1989 40% of maintenance work was thus exempted, along with so-called emergency work. The latter concession was on the grounds that, not only was it impractical to provide emergency cover using a contractor, but also that the nature of the work made reimbursement difficult, other than through daywork contracts.

Under the legislation DLOs were also able to compete for work outside their own local authority for other public bodies. Under the original Act, DLOs employing less than 30 were exempted but this was reduced to 15 in 1988, thus bringing many more under the legislation. Also in 1989, the exemption on the work below £10 000 was removed. At the same time local authorities were prohibited from renewing maintenance contracts with their own DLO, without first going out to tender.

There has been an on-going debate as to what constitutes emergency work, as there was a strong feeling that much of this exempted work was of a regularly occurring nature, which could be the subject of an item on a schedule of rates and tendered for on an open competitive basis.

In the Local Government Act of 1988, there was a tightening up of the rules relating to competition. This Act requires authorities to seek tenders from at least three non-local authority bodies, and prevents them from excluding any contractor from a tender list for non-commercial reasons.

Within the maintenance sector, it has been the case for a considerable period of time that the majority of the work that DLOs undertake comes from housing. In 1988, the figure was as high as 80% in some areas. As a result of changes in the law, the extent of local authority estates is diminishing. Social housing reforms now permit local authority tenants to choose a new landlord, whilst, in education, many schools are opting out of local authority control by taking up grant maintained status. These changes are thus reducing the preferred market for DLOs.

DLO performance

Direct labour has traditionally proved more popular for maintenance and repair work than for new-build work, and in a study of 65 DLOs immediately prior to the legislation all were involved in maintenance work, whilst only half were engaged in new-build work[1]. This analysis also showed that Labour controlled councils appeared to allot proportionately more work to direct labour, which added fuel to the ideological arguments against DLOs.

The Audit Commission[2] concluded that DLOs are market leaders in maintenance work, because of their size and approach to maintenance work and other work of a small jobbing nature. Their figures indicate that many are, by any measure, large organisations, with 9% having an annual turnover in excess of £10 million (1988 prices).

DLOs have increased their turnover since the introduction of the legislation, but their market share has reduced. Most of the fall in market share occurred immediately following legislation, after which it appeared to stabilise. An increase in the use of planned maintenance may be part of the reason for a drop in market share, as this type of work lends itself more readily to contracting out.

Since the introduction of the legislation, no DLO has been forced by the Secretary of State to cease executing maintenance work, although four have been ordered to cease trading for new work. This is not say that they have all matched the profit criterion; in 1987–88 18 out of 294 either failed to do so or made a loss.

This good performance may be explained by increased efficiency of the organisations in the execution of their work or, perhaps, to the surprise of many, due to the fact that they were not as inefficient prior to the legislation as had been claimed by many of their critics. The attitude of the parent authority may also be of some significance. The growing use of direct labour prior to the legislation suggests a strongly perceived need by some authorities for the service they provide, and led to a determination to retain them.

There are a number of ways in which help in this respect may be provided by the parent authority.

(1) The use of creative accounting, such as in the way depreciation is calculated, the valuing of work in progress, and stock valuations.

(2) Initially, in preparation of shortlists for maintenance work, documentation could be used to help give a DLO an edge competitively, although the Secretary of State has since taken action to try and eliminate this approach.

(3) By the local authority adopting a liberal attitude to claims from its direct labour department, thus permitting it to tender very competitively.

It would appear, from the limited data available, that the major fears of DLOs have failed to materialise, but that the legislation, whatever its intentions, has concentrated the minds of local authorities in the way in which they execute maintenance work. Direct comparisons of performance are difficult, as there are no comparable figures prior to the legislation.

The arguments for and against direct labour organisations

Arguments about the benefits of direct labour in local authorities have tended to divide into:

❑ Ideological
❑ A debate concerned with financial, managerial and technical performance

It is the latter of these that is of concern here, and it is perceived performance that is the prime consideration. The major argument has been that the DLO has greater flexibility for carrying out maintenance work, particularly emergency items, and that, by specialising in this type of work the DLO acquires the essential experience, both at management and operative level. Following on from this is an argument to do with the sensitivity of much of their operation, such as, for example, council house maintenance. The very real political environment within which local authority maintenance is carried out requires particular methods of management and execution. Indeed, it was felt at one time that many private contractors might be reluctant to tender for work in such a context, and this may still prove to be right if market conditions change.

All of these arguments may have been true at the time of the legislation, in that there were comparatively few private contractors with the relevant experience. Supporters of the legislation will point out, with some justification, that the legislation has changed this, and thus produced a more healthy, competitive environment in the repair and maintenance sector.

It is certainly true that there are now more contractors involved in maintenance operations, with firmly established experience. The quality of the competition does, however, depend on general market conditions. The profit potential of maintenance work appears to be less than that for new-build, although this tends to be dependent on the general level of construction activity. In a very active market the local authority, in attempting to get reasonable quotes for maintenance work, which has to be executed under tight budget constraints, may be competing on very unfavourable terms with new-build. This was one of the reasons for the growth of DLOs, and it still provides a continuing reason for their existence. In terms of cost, therefore, there may be compelling reasons for keeping a DLO because of the buffer it is able to provide.

In assessing performance, cost is not the only factor to be considered, and quality of service is an essential requirement. This is true in all sectors, but within local authorities, this has to be viewed in a social and political context.

There is also a strong technical side to this argument, in that efficient maintenance increases life cycles, and this, in the long run, produces an overall saving in expenditure. The increasing trend to planned maintenance programmes, on the other hand, makes it increasingly possible for a contractor to compete with a DLO in terms of quality.

For emergency repair work, direct labour is always likely to have an edge, although there are problems associated with this. The first concerns the definition of emergency work and what can be realistically put into a planned programme. There is no doubt that this is vulnerable to some manipulation, and can be used by local authorities, with a strong political will, to retain a DLO in difficult operating circumstances. The other difficulty is that the most carefully produced planned programmes are subject to the vagaries of local government finance, and to the incidence of essential emergency work. This causes complications in contractual arrangements.

Critics of direct labour have consistently argued that, by its nature, it is protected from the forces of competition, and will thus be riddled with waste, inefficiency, overstaffing and restrictive practices. There is little evidence to suggest that this is now true.

Supporters of DLOs, on the other hand, point to their excellent records with respect to safety, health and welfare, and training. This latter point does seem to be important, with increasing concern about training levels and standards within the industry as a whole. Local authorities do have a strong commitment and an excellent record in training, and this is particularly true with respect to maintenance oriented staff and operatives.

These arguments, which relate to public sector maintenance, identify many of the issues to be considered by any estate or building manager. However, because of the context within which public sector estate holders have to work, the decision making process may be somewhat distorted.

In most instances, in practice, both direct and contract labour will be used, and acceptance of the rationale for this is important for all decision makers. The real decision making should especially be concerned with the most appropriate mode of execution, having regard to both context and corporate objectives.

For a public sector organisation, both context and objectives will be influenced by social and political factors, as well as economic ones. In the private sector commercial imperatives may be the dominant force. In both cases perceptions and notions of accountability should not be ignored, notwithstanding the differing scenarios within which each of these exist.

Direct versus contract labour

A simple, if imperfect, division of maintenance work is into emergency or unpredictable work, and planned or predictive maintenance. It can be assumed that the latter lends itself to the use of contracting, and this may or may not be to a DLO. The criterion for choice will almost inevitably be on price, although even with competitive tendering there will be some scope to select contractors to go on to a list on the basis of the quality of service they are able to provide, based perhaps on historical data.

Emergency or non-predictive maintenance presents more problems in that, although it may provide a powerful argument for maintaining a pool of direct labour, some of it may be contracted out.

The major debate has to be about the proportion of work to be allocated to each, which will normally be a question of deciding on the size of the direct labour pool (figure 8.1), and the setting up of appropriate contractual arrangements to provide the client body with a suitable degree of control and accountability. The latter may also be necessary for direct labour, depending on the type of accounting conventions to be adopted, such as in the case of a local authority DLO. Additionally, a payment mechanism will need to be put in place if direct labour is to execute emergency work on a non-competitive basis.

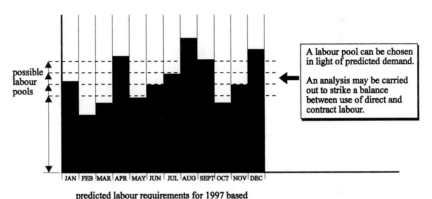

possible labour pools

A labour pool can be chosen in light of predicted demand.

An analysis may be carried out to strike a balance between use of direct and contract labour.

JAN FEB MAR APR MAY JUN JUL AUG SEPT OCT NOV DEC

predicted labour requirements for 1997 based on historical data and programmed work

Figure 8.1 Choosing on appropriate labour pool.

A problem that then arises is that a contractor, whether private or a DLO, in order to ensure its own survival, has to make commercial decisions. If it is to execute emergency repair work with the quality of service required by the client department it has to ensure that it has resources available to it, e.g. a labour pool. This may require a pool of labour in excess of that normally required by day-to-day operations, with the attendant costs this imposes. Indeed, one of the arguments used by defenders of DLOs against the perception that they were inefficient, derives from this principal. There is no doubt that there is a high cost attached to rapid response, and this is the strongest argument for the implementation of rigorous planned maintenance programmes.

Within the public sector, it is, therefore, the case that much of the decision making about who does maintenance lies beyond the client department. Within the private sector, the decision making is almost certainly going to be made on a commercial basis, with clearly defined response requirements probably only for specific parts of the building. It is also likely to be the case that any direct labour used is employed within the maintenance department, i.e. it will be part of the maintenance cost centre. The position of the maintenance department within the organisation will vary, as will the proportions of work executed in-house and by contract labour. This may or may not be the result of a logical decision making process.

Organisation of maintenance departments

Whilst the relationship between a maintenance department and the rest of the corporate body can be extremely variable, it is possible to identify common elements or operations that will exist in the departments themselves. These include:

❑ Identification of maintenance work, both planned and emergency, which can be called work input
❑ Instructions for the work have to be transmitted to the work team
❑ Execution, supervision, approval and valuation of the work
❑ Authorisation and making payment
❑ A contractual framework
❑ An accounting context
❑ A feedback system, the sophistication of which is variable

These can be represented in the maintenance system as illustrated in figure 8.2, which is based on Wiener's view of an organisation as an adaptive system, entirely dependent on measurement and correction through information feedback[3].

Maintenance organisations can be broadly encapsulated into two types of organisation:

❑ Centralised
❑ Decentralised

There are significant differences in these models, relating to operational aspects, but one of the most significant features relates to the size of organisation required at the centre. This will vary, not only with the policy adopted by the parent organisation, but also through a range of factors directly related to operational matters:

❑ The nature of the building stock
❑ Volume, timing and diversity of the workload
❑ The complexity of the stock in technical terms
❑ Geographical and topographical factors

- ❑ Restrictions on the timing of work
- ❑ Whether work is to be executed by contract or direct labour
- ❑ The level of expertise of the work force, and the extent to which non-operational tasks, such as routine inspections, are delegated to the people on the job
- ❑ How maintenance is defined in the organisation, such as the degree of involvement of the maintenance department in minor capital works

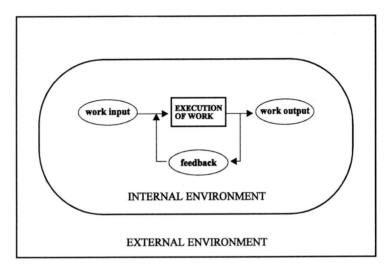

Figure 8.2 Maintenance as an adaptive system.

The two examples shown in figures 8.3 and 8.4 allow for the possibility of work by contract or direct labour, although at the detailed level, many management functions will be affected by the balance between these two.

Initiation of maintenance requests

The key document for initiating individual maintenance items is the work order, or the job ticket, irrespective of whether this is issued to direct or contract labour. It must be remembered that this document, as well as initiating work, may be an essential component of the reporting and recording mechanism. A great number of variations exist, but in general they are designed following similar principles.

A range of work orders can be used within an organisation to suit varying requirements, but this approach is difficult to justify. It is better to simplify the work order in order to assure uniformity, and recover feedback and control information by other means. For example, it should be possible to utilise the same basic work order for planned and

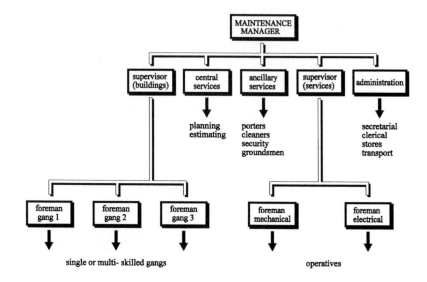

Figure 8.3 Centralised or functional maintenance organisation.

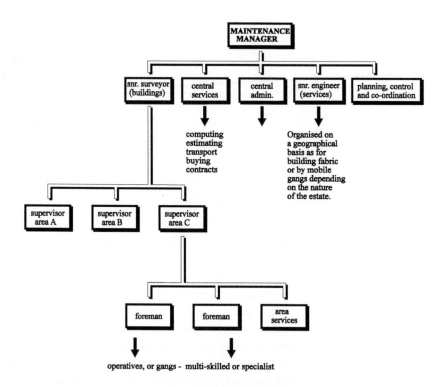

Figure 8.4 Decentralised or geographical maintenance organisation.

unplanned work and, particularly in cases where maintenance gangs execute a mix of the two in any day, this is a sensible approach. The following information should be included as a minimum.

(1) A job number and the date of issue.
(2) Address of the property or an appropriate location code.
(3) A work classification code if one is in use.
(4) The location of the defect or service required.
(5) As accurate a description of the work as possible, and the use of standard descriptions may be useful here, particularly if they can be coded in some way, perhaps linked to a classification code.
(6) Times and means of access, accompanied, perhaps, by a trigger code to indicate when permission to work is required.
(7) For non-planned items a priority coding will be necessary to allow teams to plan their work. In practice a four point approach seems to be the most sensible:

   (i)   Emergency work for which reasonable requests would receive at least essential attention within 24 hours.
   (ii) Urgent work which does not constitute a danger, but which should reasonably be completed in a week.
   (iii) Normal work, which will probably account for the major portion of the contingency allowance for non-planned work, and which may be included into a medium term programme of work, perhaps on a three-monthly basis.
   (iv) Standby work which is of low priority and can be held in reserve and fed into a planned programme at strategic intervals, in some cases, perhaps, to take up any slack in programmes.
   In addition, a 'P' code can be entered against a planned item.

(8) Estimated labour hours can be added to the job ticket, along with a space for the entry of the actual hours taken. This is facilitated if there is a library of standard jobs, coded, with a labour usage against it. This may not only be important from the point of view of control, but can also assist work planning. If a bonus system is operating, there may also be a target time entered, which may not be the same as the estimated time. If contract labour is being used, then the information requirements on the order will reflect the contractual arrangement entered into.

   The entry of hours taken needs careful consideration, as there will inevitably be a proportion of non-productive time, for example due to travelling. This should be separated out for analysis and feedback purposes, and for comparison of actual response times against priority codes.

(9)   Estimated materials quantities may be entered. However, in all but simple cases, this may be expecting too much of the system. A separate materials management system may be indicated.

Figure 8.5 illustrates the layout of a works order produced by the FrontLine™ maintenance management system. In common with many other computerised systems, FrontLine™ provides a considerable customisation facility within the package to produce simplified versions, if appropriate. This example of a comprehensive pro-forma, however, demonstrates the potential for basing management systems around electronic databases. The works order is generated by the system and permits a high level of analysis to take place.

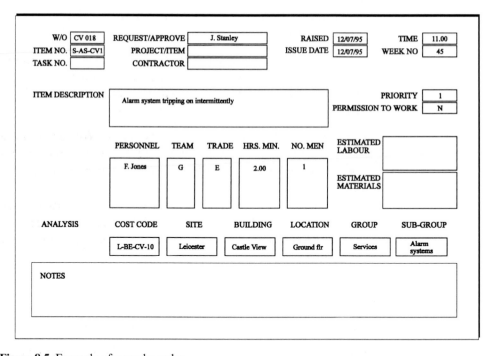

Figure 8.5 Example of a works order.

Pre-execution functions

Once a job has been initiated there are certain management functions that need to take place before execution and afterwards; and before the works order is passed for execution any of the following may be necessary:

(1)   A check to ensure that the item has not been initiated by any other means, for example, an occupant request is not already in a planned programme.

(2)   A check to ensure that a request is valid, i.e. expenditure is justified, and there is no other conflict with work already initiated.

(3)   A decision made with respect to a priority rating. This is a difficult area, and it is convenient if this can be decided by the person receiving the request, so as to hasten the speed with which the job enters the system. The extent to which this is possible depends on organisational priorities.

(4)   A physical inspection of the defect or service required may be necessary for any of the following reasons:

- ❏   To estimate costs or resources required
- ❏   To arrange access
- ❏   To verify information for entry in the works order
- ❏   For diagnostic purposes

The general principle adopted will tend to be to utilise exception principles, as one-off inspections add greatly to the cost of maintenance work. If work is of a low priority, however, such an inspection can be integrated into work execution.

The details of the process will obviously depend on the nature of the organisation and its building stock. The procedure outlined in figure 8.6 is a typical one, and again indicates the use of a database, which is a useful tool in checking for duplication.

Overlying all these requirements, is the necessity to consider user requests as being very much the customer interface, and in terms of occupational efficiency and the performance of users, it is unwise to underestimate physiological and psychological influences.

Post-execution functions

Following completion of the work any of the following may take place:

(1)   An inspection of the work, to check that it satisfies requirements and to authorise payment. For contract labour the contract clauses will lay down certain requirements, as it is impractical to inspect every job in this way. In the case of contract labour, there has to be an implicit understanding that contractors will seek to provide a satisfactory service in order to retain the work.

In the case of direct labour, it is a function of the maintenance department to establish a set of norms in terms of service expectations. It may have been decided, as a matter of policy, to inspect a range of jobs at random. It is desirable, however, that properly organised user feedback is introduced to provide sensible satisfaction intelligence.

(2)   Following a certified completion of the job, the property record will need to be updated.

(3)   The cost will be entered into an accounting system, either external to the maintenance department and/or internal, depending on the way in which costs are being recorded and analysed.

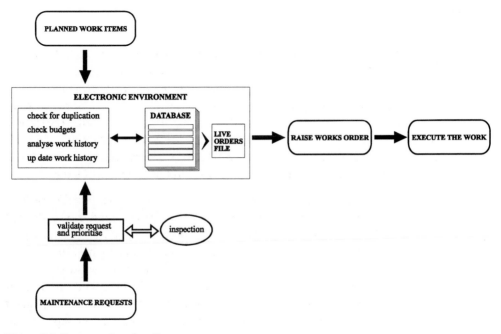

Figure 8.6 Pre-execution functions.

Figure 8.7 illustrates a typical procedure that might operate under an electronically driven database system.

Execution of work

There are a number of particular problems associated with the execution of maintenance work, which tend to revolve around the characteristic diversity of maintenance operations, the small scale nature of individual items of work and the need for the attendance of a number of different trades. The problem of ensuring a proper sequence of trades, and balancing the relative numbers of each trade, is a productivity issue

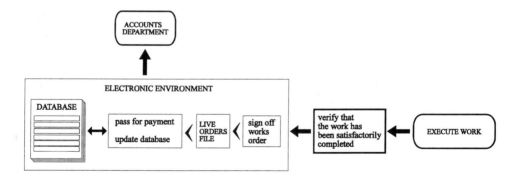

Figure 8.7 Post execution functions.

throughout the whole industry. As the size of an individual task reduces, there is a tendency for this problem to be exacerbated.

It may be assumed that planned maintenance programmes will be executed in a systematic way, and that reasonable planning of the sequence of operations and the balancing of the various trades is a realistic proposition, whether it be contract or direct labour.

There are several ways in which the labour force can be organised, and these are of equal relevance to both direct and contract labour.

(1)   A single group of operatives, consisting of one specific skill or specialism, may carry out a planned programme over a defined geographical area. An example of this might be a planned programme of flat roof replacement, carried out by an external contractor. This is called a *homogeneous contained* operation.

(2)   A problem occurs, however, when such an operation requires additional work, of a supportive nature, by another trade. A good example might be a planned programme of plumbing replacement work which requires a making good and decorating operation following its completion. Under some circumstances it may be possible to attached a multi-skilled building operative to the team. This needs careful consideration to ensure a correct balance between the two types of work, and assumes the availability of such an operative. Some of the most obvious and commonly planned routine programmes may encounter this problem. For example, redecoration may be accompanied by associated joinery or masonry work. If this has been identified prior to commencement, some planning can take place, otherwise the act of redecoration will identify the need for work which was not foreseen. This is termed a *homogeneous uncontained* operation.

(3)   Other items of planned work will, by their very nature, involve a

number of trades, and the planning problem will be more difficult, but still realistically possible. An example of this might be a window replacement programme that requires bricklayers, carpenters, glaziers and decorators. This is a *heterogeneous* operation. If each trade requires the same amount of time on each unit then planning may be straightforward. Unfortunately this is rarely the case, so that a balancing operation will be necessary.

(4) There are other planned items that do not fit into any of these categories, as they are not necessarily repetitive in nature. These will largely consist of items uncovered by planned inspections and low priority maintenance requests, and will clearly need to be planned in rather a different way. These may be termed *semi-plannable* items and, due to the organisational consequences, need separate consideration.

Ignoring these semi-plannable items for the time being, depending on the geographical dispersion of the estate, gangs can be organised on an area basis or by specialism.

(1) Homogeneous contained operations are probably better organised on the basis of a specialised gang operating on an estate-wide basis. Economies of scale suggest that sub-division is only realistic on very large estates.

(2) The organisation of homogeneous uncontained operations will depend on the amount of support work necessary. It has to be borne in mind here that the percentage of non-productive time involved in a making good operation may be very high, so that, when this is only of a very minor nature, there may be a case for regionalisation. Where the support work becomes more extensive, it may become realistic to have support operatives who travel rather further.

(3) The organisation of heterogeneous operations on a planned basis can be approached in two ways. Single trades may be rotated through the programme of work on an individual basis, or, a balanced multi-trade gang may be set up. There are many advocates of the latter, on the basis that group relationships will form, and that benefits may flow from this. Ultimately, however, the viability of setting up such a gang depends rather on the nature of the operation. Multi-trade gangs are more likely to be organised on an area basis, and to be responsible for more than one operation. Highly specialised gangs, on the other hand, will probably operate on a wider geographical basis.

Much more troublesome is unplanned work, of which several types, presenting varying levels of organisational problem, can be identified.

(1) Items of work may be identified from a programme of planned inspection. Depending on the priority level, it may be possible to insert this work into a planned programme; i.e. it can be termed semi-plannable.

(2) Items may be initiated by a user request which, depending on the level of priority, can become semi-plannable or, if very low priority, part of a fully planned longer term programme. Many items from this source, though, will inevitably fall into the category of unplannable.

(3) Additional items uncovered by a normal planned programme present a large dilemma. They can be divided into two categories.

  ❑ Those whose rectification may be a pre-requisite for the completion of the primary maintenance operation, so that they require immediate treatment. These give rise to organisational problems, sometimes of a complex nature.
  ❑ Those treated like a user request, whose execution will depend on the priority attributed to them. It may, of course, be deemed appropriate to execute them immediately, purely on the grounds of efficiency, provided the planning of the primary item allows.

There are many items which will disrupt the careful organisation of work, and it can be seen that the distinction between planned and unplanned maintenance may become a little blurred at the execution level.

Items that have been termed semi-plannable will not present an immediate problem, in that they will be inserted into a future programme in a controlled way. The main cost that has to be borne is the disruption to planning, and this emphasises the need for programmes to be dynamic and flexible.

Of greatest difficulty to organise are those items requiring immediate attention, and the two categories that have been identified above merit separate consideration. Those immediate items, uncovered during the execution of a planned operation, can be exceedingly disruptive, and, even with the most careful programme of inspection, can never be completely eliminated. As well as the obvious repercussions for financial planning, they will inevitably disrupt the operation of a gang, often necessitate the calling on to the job of another trade, and sometimes force the temporary movement of the gang to another operation, possibly on another site.

The amount of disruption this all causes depends on the availability of another task appropriate to the gang concerned, and the response time of the additional labour involved. This highlights the benefits of a loose-fit style of management[4].

Urgent maintenance requests, not a pre-requisite for the current operation, can be dealt with in a number of ways.

(1)   The setting up of multi-trade gangs that are purely executing planned items may not be realistic and is not, in any case, always the most effective way of organising the work. It is common practice for multi-trade gangs to receive a daily or weekly pro-gramme that is a mixture of planned and unplanned work. This provides one useful mechanism for the treatment of urgent repairs originating from users or in the field (figure 8.8). Under such a system, maintenance requests are logged and ordered according to priority rating, and allocated to each maintenance team according to their planned whereabouts and work load. Teams of this nature are likely to cover an area of an estate.

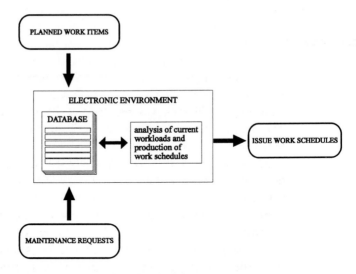

Figure 8.8 Generation of work schedules.

(2)   Maintenance teams may be very small, and in some cases only one or two multi-skilled persons can be of great value. There are advantages to be gained if these very small teams are based in a very localised area, as they can build up good working relations with users. This is an approach advocated in housing management, where the local works office not only executes work, but receives

maintenance requests and initiates work under rather looser central control. It should be remembered, however, that some of the thinking behind this is sociological in nature, rather than based on maintenance efficiency considerations.

(3)   There will be a proportion of maintenance operatives who will solely execute emergency work, and these will almost certainly, except on very small estates, be responsible for a localised area.

The allocation of manpower solely to emergency work is a difficult process, and depends on being able to reasonably predict the volume of work and the response times required. There has been some experimentation with this type of work, particularly in housing maintenance, where small multi-skilled teams are employed, based on one estate with the objective of not only executing maintenance work but fostering good landlord/tenant relationships. This approach, it is hoped, will reduce vandalism and, hence, reduce the number of maintenance requests. The economics of such small scale operations are difficult to assess, and cannot be judged on raw maintenance efficiency criteria alone.

The maintenance operative

Maintenance operations are very labour intensive and, therefore, the greatest potential for economising on the execution of the work lies with efficient use of labour. Studies have shown that, for painting and decorating, labour accounts for 85% of the prime cost, and for general repair work[5] as much as 65%.

The scope for improvement is largely to do with the efficient organisation of the work, as mechanisation has relatively little potential. Private contractors who engage in maintenance work are also generally small in nature and reluctant to invest, even when there is potential.

Somewhere in the region of 40% of the construction industry's work force is employed in repair and maintenance. Amongst some trades, the proportion involved in maintenance work is very high, for example about 75% of all painters and decorators. The training of these operatives is largely along traditional lines, with very little specific to maintenance. A large volume of maintenance work is carried out within the traditional trade divisions, and many small contractors use the same labour for both new-build and repair and maintenance.

It is quite normal for maintenance operatives to work in small gangs, made up of a number of trades, and there is evidence of a far greater blurring of the divisions between trades than occurs in new-build work[6]. Furthermore, there are clear indications that the use of multi-skilled personnel can bestow large advantages in terms of flexibility of working. There are obstacles to these practices, and exceptions are more prevalent in Scotland and the north of England, and within local authorities. The assumption here is that trade union influence acts to more strongly reinforce demarcation lines.

Rigid trade demarcations clearly hinder the provision of an efficient and effective maintenance service and concern about this on an industry-wide basis has been expressed in a series of reports, dating back to the end of World War II. For example, the Phelps Brown Report[7] contained evidence commenting that an important part could be played in controlling the rising cost of maintenance by the encouragement of multi-craft skills.

Of major concern is the steady de-skilling of new construction, with the increasing tendency for buildings to become a series of assembly operations. This has contributed to the decline in training, in terms of volume, and also in the nature of the essential experience to which a trainee can become exposed. As the level of skill required to execute work on existing buildings is generally higher than that required for the average new-build project, there is mounting concern that the supply of such skilled labour will diminish, as the effects of reducing training programmes work their way into the system.

The labour force in maintenance tends to be older than that engaged in new-build, and there are logical reasons behind this. The proportion of operatives over 40 years of age has been estimated as 58% for maintenance and only 32% for new construction[8]. A newly qualified tradesman has high earning potential in his early twenties, and this potential is more likely to be realised in new construction. However, as he gets older, and perhaps undertakes additional family responsibilities, this earning potential is counterbalanced by the uncertain nature of his work, and he is likely to seek more security elsewhere in the industry. Maintenance work has tended to provide one of these more stable positions. This was, of course, particularly true in local authority DLOs, and although the situation has changed somewhat, maintenance output remains rather more stable than does new construction.

It has been argued that the varied nature of maintenance work provides a higher level of personal satisfaction, particularly as there is a close relationship to the actual use of the building. There is a suggestion, therefore, that the maintenance operative is less motivated purely by financial reward, although it is difficult to separate these motivations from what many believe to be the disincentives of larger construction sites.

In proposing the need for a multi-skilled operative, it is necessary to be aware not only of practical difficulties, but also the need to determine what basic skills are required. Maintenance involves such a wide range of activity that it is difficult to produce the perfect maintenance technician. The need for specialists will always exist, but many of these skills may be rather more contained. There is a strong case for arguing that the real provision of the type of skills required will come through experience and on the job training, rather than in a classroom. Individual craft courses can, however, do more to educate trainees in the broad nature of all construction work, so that their role can be seen in context.

Many of the demarcation problems that occur in maintenance are related to the need to execute support work. Making-good exercises are extremely common, and the skill level required for them is not necessarily high. The average carpenter is almost certainly quite capable of making good a small area of brickwork after fitting a new window[9].

Productivity

There are a number of ways in which productivity can be measured, although it is a notoriously difficult subject on an industry-wide basis, as the interdependence of gangs and trades, and other characteristics of construction operations always make it difficult to pin down responsibility[10]. Maintenance work provides even more problems than new construction in this respect, and raw productivity data, no matter how accurate it may be, is by no means a firm indicator of the health or otherwise of a maintenance department.

In general, productivity depends on the work rate and the proportion of attendance time spent on productive work, and there are a number of measures that may give an indication of labour productivity.

(1) Gross output per operative
This is given by the total cost of wages, materials and overheads, divided by the number of operatives employed. It gives a simple crude evaluation over the organisation as a whole, but is of little value for inter-trade comparisons.

The ratio is also vulnerable to overheads, in that an increase in overheads suggests an increase in productivity. The work output measure may also be affected by political considerations, or budgetary constraints. A cut in budget that restricts output, reduces productivity as measured by this method.

(2) Value of material used

This can be expressed as a total figure, over a period, to permit comparisons of output over time and/or in terms of material output per gang. For the purposes of comparing gang performance, it is of little value, as the material usage of each gang varies due to factors other than output. The figure can also be distorted by high wastage levels. Its main use is for evaluating the general level of output over time, in a similar fashion to the way brick stocks are used for the industry in general.

(3) The rate of production of repetitive jobs

For repetitive jobs, it may be possible to do simple comparisons, either in terms of the number of jobs executed in a given time, or by an analysis of the times taken to complete jobs. There are obvious limitations to the appropriateness of this type of measure.

(4) Comparisons with programmed or estimated hours

There are several variations on this theme, which have close links with work study techniques. The simplest approach is to divide the estimated time by the actual time taken. A more sophisticated variation is the development of a set of data, which expresses the work content in terms of a number of standard hours. These are defined as the amount of work that can be performed in one hour by a standard skilled and motivated worker. This information can be determined from work study measurements, or from past output records, if they are available. There are then a number of ratios that can be calculated.

For example, the performance factor is equal to the standard hours of work produced divided by the actual hours of work expended. Other measures include cost per standard hour and gross cost of an hour's actual work.

In all these methods, there is a reliance on the derivation of either estimated times for executing work, or a standard hour, both of which present obvious difficulties. Within this is also some assumption that one job will always be like another and, whilst some element of repetition can reasonably be assumed, this is by no means as simple for maintenance work as would be expected. There will also always be items of work for which standard output targets cannot be derived, and their presence alone will distort productivity measurements.

There are other factors that can be evaluated. For example, one indicator of the effectiveness of a planned policy is to identify, for operatives or gangs, the number of hours spent directly on planned work, and compare this with the total hours spent. Interpretation of the results needs to be cautious, but this may be a useful management indicator, provided the full range of possible reasons for poor performance are investigated.

It is also possible to determine the amount of lost time incurred by a gang, and again this might indicate management shortcomings and problems with work allocation procedures, as well as inefficient operative practices. If a proper reporting procedure has been instituted, it should be possible to classify lost time into any of the following causes:

- Waiting for instructions
- Waiting for materials
- Waiting for access
- Travelling time
- Other

The Audit Commission[11] advocate the determination of the percentage productive time. They comment that it would be normal to expect this figure to be about 80%, but add that its measurement is subject to a great deal of variation, and this makes for considerable difficulty in comparing organisations.

None of these measures can be taken as particularly useful in absolute terms, but they all have some value in identifying trends. The important thing is to always seek to discover the real underlying reasons for a poor performance in any respect.

Raw productivity in maintenance work will always tend to be lower than for new-build work due to a number of factors:

- The small scale nature of the majority of individual jobs leads to a high overall percentage of non-productive time
- The diversity of job content means that demarcation delays can be extensive
- The low level of repetition in unplanned work
- The wide dispersal of individual sites
- The often difficult working positions
- The cost of obtaining access to areas for maintenance is a particular problem; scaffolding, for example, can cost many times more than the actual cost of the repair itself
- The ancillary work, such as making good, associated with a maintenance operation
- The sometimes high cost of obtaining materials to match existing work in older buildings

Due to all these factors, there is increased pressure on the efficiency of planning, the adoption of proper inspection routines to minimise disruption to planned programmes, and the adoption of appropriate working methods. A number of Audit Commission reports have been frankly critical of the general quality of estate management in the public sector on these grounds.

Other performance indicators

There are a number of other performance measures that may be better indicators of the effectiveness of the whole maintenance operation, rather than just its execution.

(1) Delay and response times

Work orders should indicate the level of urgency of an unplanned repair, and/or the programmed date of a planned repair. One of the major objectives of planning is to provide a control mechanism, so comparison of actual progress with planned is an essential feature. Caution needs to be exercised on the action to be taken if progress is behind schedule, as there are a multiplicity of factors that may quite reasonably disrupt planned maintenance.

For emergency work, evaluations can be carried out, either by determining the average response times and comparing them with those required, or by determining numbers of backlog jobs and analysing them.

(2) User satisfaction

A simple method of evaluating user satisfaction is by the recording of customer complaints, although the interpretation of these is somewhat difficult. The level of customer complaints may well be influenced by factors other than the quality of maintenance.

Housing associations regularly use the number of voided units as an indicator of their performance overall, and indeed set them as a management objective. Good maintenance is again, however, only one factor in this.

(3) Costs of failure or breakdown

In many commercial and industrial organisations this is the prime measure of the performance of maintenance, although factors outside the maintenance department may be influential. It is important to underline the consequences of an unenlightened corporate policy, and to emphasise the need for any performance indicator to be evaluated in the context of the whole system.

Motivation and incentives

The need to motivate a maintenance work force is just as much a part of management responsibility as it is anywhere else in industry, and there is a large body of literature available that discusses theories and approaches.

There are a number of factors that may work in favour of maintenance operations in this respect, when viewed in comparison with construction industry alternatives. Unfortunately, many of these factors are negative rather than positive, and reflect a generally poor view of the industry as a whole. The problem remains one of attracting good calibre people into the industry and retaining them.

Financially-based incentive schemes still have a place, and are an obvious way of trying to improve productivity. There are a number from which to choose.

(1) Piecework

The oldest type of incentive scheme to be used is the piecework approach, which in its simplest form operates on the principle of a target, in the form of a lump sum, payable for completion of a given task, irrespective of how long it takes. There is, therefore, no guaranteed wage, and payment is made for productive work only. The real incentive is to get a job done as quickly as possible. However, the obvious danger is that a poor finished product results, possibly executed using unsafe working practices.

The method also relies heavily on the setting of realistic targets, and some studies have shown that it leads to high labour turnover, as work gangs have loyalty only to whoever pays the best price. Market conditions are a major factor, and militate strongly against its application to maintenance work.

(2) Spot bonuses

These are additional sums of money, paid over and above the normal remuneration. They tend to be applied in one-off cases, where a particular service is required, such as the rapid completion of a particular item of work.

(3) Work study based schemes

A range of work study based schemes have been developed to try and counteract the deficiencies of piecework type approaches. Work study uses the British Standard Institute rating scale, as shown in figure 8.9. In this scale 75 is taken as normal performance. Work measurement is then

SCALE POINT	DESCRIPTION	COMPARABLE WALKING SPEED M.P.H.
50	Very slow, clumsy, fumbling movements, operator appears half asleep, with no interest in the job.	2
75 NORMAL PERFORMANCE	Steady, deliberate, unhurried performance as of a worker not on piecework but under proper supervision. Looks slow but time is not being intentionally wasted whilst under observation.	3
100 STANDARD PERFORMANCE	Brisk, businesslike performance, as of an average trained worker on piecework. Necessary standard of quality and accuracy achieved with confidence.	4
125	Very fast, operator exhibits a high degree of assurance, dexterity and coordination of movement well above that of an average trained worker.	5
150	Exceptionally fast, requires intense effort and concentration and is unlikely to be kept up for long periods. A virtuoso performance only achieved by a few outstanding workers.	6

Figure 8.9 BSI rating scale.

used, on a discrete operation or task, to determine a measured or observed time. At the same time the operative's speed of working is assessed, in accordance with figure 8.9, and a basic time is derived from the following formula:

$$\text{Basic time} = \frac{\text{observed time} \times \text{rated speed of working}}{100}$$

If the rated speed of working is assessed at 100, then the basic time is the same as the measured time. If a normal performance factor of 75 is used then the basic time is 75% of the observed time.

Allowances may then be made to the basic time to take account of working difficulties and other factors to produce, what is called, a standard time. This allows the setting of what is termed a target time for an operation, and there are then several ways in which bonus may be determined.

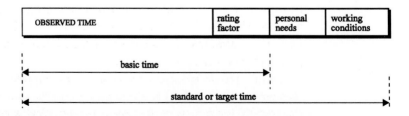

Figure 8.10 Determination of standard time.

The earliest approach was to use what is called a directly proportional scheme, where target times are issued at the normal 75 rating in terms of standard times. Operative performance is then assessed, and a bonus paid over and above the basic rate in direct proportion to performance. Above the 75 performance level, bonus increases in direct proportion to performance, as shown in figure 8.11. There is normally a cut-off at the 125 performance level to ensure good working practices.

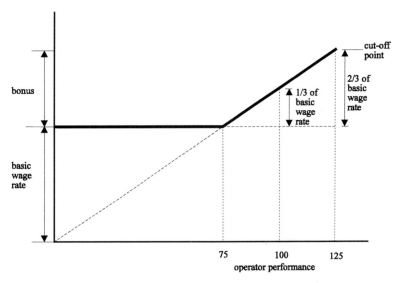

Figure 8.11 Direct proportional incentive scheme.

A variation on this is to set target times at the standard performance level of 75, and then measure the time saved on an operation by an operative. Standard performance is achieved when 60 standard minutes worth of work is produced in 60 elapsed minutes. If an operative produces 100 normal minutes worth of work in an elapsed hour, then they have made a saving of 40 minutes, for which they are paid. Their bonus rate would be 40/60 or 2/3 of the basic wage rate. This can be put another way by saying that producing 100 minutes worth of work in 60 minutes is equivalent to a performance of 125, and, looking at figure 8.11, the same result is achieved. The second of these two methods has been advocated, on the basis that operatives can understand it more easily when they are paid their bonus as a time saving.

Not all schemes pay the full saving to the operative. Quite reasonably, management often argues that good performance is not just made possible by the efforts of the operative, and that some of the savings should be retained in the organisation, or perhaps used to reward good management. This results in what is called a geared scheme as illustrated in figure 8.12.

A further factor to be considered is that all construction work, and maintenance in particular, is extremely variable in nature, and if a simple bonus system is used then earnings can fluctuate severely, often through no fault of management or operative. One method of stabilising earnings is to calculate operative performance over a longer period than a week, so that in effect bonus payments become averaged. It is also possible to use a non-proportional approach, as indicated in fig 8.13. In this example bonus payments start at 50 performance and rise rapidly to 25% of basic pay at standard performance, but at a diminishing rate thereafter.

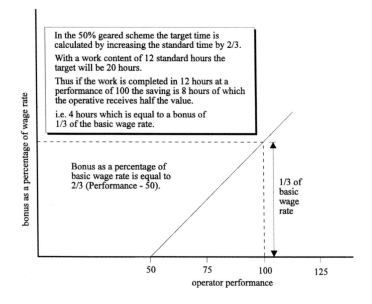

Figure 8.12 Geared incentive scheme.

For all these types of incentive scheme, targets have to be reviewed at regular intervals, as there is a tendency for the time taken to execute a job to decrease over time. This is referred to as 'slippage'. For this, and a number of other reasons, in recent years, local authorities, in particular, have moved away from this approach in favour of other methods.

(4) Allowed time schemes
These schemes use previous work study data, but express target times at performance rates between 50 and 75, with non-productive time allow-

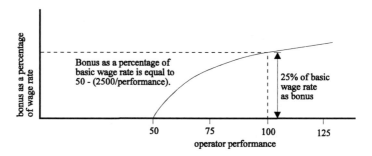

Figure 8.13 Non-linear geared incentive scheme.

ances, to produce what is called an allowed hour of work. Earnings are calculated by multiplying the allowed hours by an agreed rate per allowed hour.

(5) Tender led schemes

These, in general, work on very similar principles to the allowed hour approach, but with rates from tenders used as the basis. A major disadvantage is that earnings within an organisation can become extremely variable and lead to discontent.

(6) Indirect schemes

There are a number of indirect incentive schemes that can be used.

(1) Profit sharing
  ❏ Immediate profit sharing is where a percentage of the gross profit at the end of a job is shared out pro-rata to the operatives. This provides a good incentive at the work face.
  ❏ Delayed profit sharing is where profits are shared out on a company wide basis, annually or bi-annually. This approach can be more useful in maintaining and promoting longer term loyalty.

(2) Merit rating
  This can be applied through a selective form of plus rates, made over and above basic rates, for operatives who have demonstrated their merit. There are a variety of merit rating schemes that may be applied, but they nearly all suffer from problems of objectivity, although some criteria such as length of service are easy to measure and reward on a consistent basis.

Supervision

There are problems associated with the supervision of construction work in general, and this is exacerbated by the characteristics of maintenance work. There has been a tendency to criticise new construction sites for not providing sufficient back-up staff on a site, whereas in the maintenance sector, and particularly where DLOs are concerned, the tendency is to comment unfavourably on the proportion of non-productive staff.

Within maintenance organisations, the roles allotted to supervisors can vary greatly but, for simplicity, it is useful to consider here the management who have direct responsibility for work execution and its control, including productivity.

The organisation models outlined earlier demonstrate the extreme approaches that can be taken but, in general, it can reasonably be assumed that the maintenance supervisor will be mobile and that, therefore, a large amount of responsibility resides at the work face, in the hands of the operatives. The relationship with them is therefore of prime importance.

At the organisation level, good, well-run, bonus schemes will help maintain progress and productivity at the work place. This can have the benefit of freeing the supervisor to play a more constructive role, in terms of developing tenant/landlord or user/occupant/owner relationships. In many cases, the supervisor's role can also become a justifiably broader one, and encompass the whole range of attributes that help ensure an effective maintenance service.

References

(1)   Lowe, J.G. (1987) The Use of Direct Labour in Local Authority maintenance contracts, In Spedding, A. (ed.) *Building Maintenance and Economics – Transactions of the Research and Development Conference on the Management and Economics of Maintenance of Built Assets*. E. and F. Spon, London.
(2)   Audit Commission for Local Authorities in England and Wales (1989) *Building Maintenance Direct Labour Organisations – A Management Handbook*. HMSO, London.
(3)   Wiener, N. (1948) *Cybernetics*. MIT.
(4)   Bishop, D. (1982) Productivity – Whose responsibility? In Brandon, P.S. (ed.) *Transactions of Building Cost Research Conference – Building Cost Techniques*. Portsmouth Polytechnic, September.
(5)   Holmes, R. (1983) *Feedback on Housing Maintenance – BMCIS Occasional paper*, July. RICS, London.

(6) Building Research Establishment (1966) *Building Operatives' Work*. HMSO, London.

(7) (1968) *Report of the Committee of Enquiry under Professor Phelps Brown into Certain Matters Concerning Labour in Building and Civil Engineering – CMD 3714*. HMSO, London.

(8) Department of the Environment, Welsh Office and Scottish Office (1993) *Housing and Construction Statistics 1983–1993*. HMSO, London.

(9) Audit Commission for Local Authorities in England and Wales (1989) *Building Maintenance Direct Labour Organisations – A Management Handbook*. HMSO, London.

(10) Bishop, D. (1982) Productivity – Whose responsibility? In Brandon, P.S. (ed.) *Transactions of Building Cost Research Conference – Building Cost Techniques*. Portsmouth Polytechnic, September.

(11) Audit Commission for Local Authorities in England and Wales (1989) *Building Maintenance Direct Labour Organisations – A Management Handbook*. HMSO, London.

Appendix 1
Statistics

In the day-to-day running of a building maintenance department a great deal of data will be generated. If this data is to be of any real value, the maintenance manager will need to know how to collect, organise, analyse, interpret and present it effectively. To do this, a basic knowledge of statistics is necessary, and to illustrate the point some basic principles are outlined in this appendix, using maintenance related data.

Collection of data

Maintenance operations, by their very nature, generate large quantities of data, much of which will be recorded, i.e. it is collected. How much information is actually collected depends on management objectives, and this will heavily influence the choice of management systems.

Even when a comprehensive data set is available, it may be considered necessary to only use a sample of it for an analysis. The selection of this sample is an important consideration, as its size and nature will influence the results and conclusions drawn from its analysis. To obtain a meaningful result, the sample would have to be large enough to be significant, and chosen so as to be representative. For a comprehensive treatment of sampling techniques reference can be made to a number of texts[1,2,3].

Presentation of data

Statistics is concerned with data of all kinds and, leaving aside the question of its analysis and interpretation, a primary requirement is to present it in the most suitable manner. Tables are the simplest way of presenting information, but there are numerous approaches possible. Figure A1.1 shows statistics for local authority DLOs in the period 1970–1980. Another example, shown in figure A1.2, is taken from a study of the maintenance costs of CLASP buildings[4].

	% of total public and private sector work				% of public sector work			
	all work	new work	repair and maintenance		new work			repair and maintenance
			housing	other	total	housing	other	other than housing
1970	10.5	3.5	17.7	37.1	6.8	6.4	6.9	54.0
1971	10.4	3.3	18.0	38.0	6.8	6.1	7.1	54.5
1972	10.4	3.2	18.2	37.7	6.8	5.6	7.4	52.8
1973	9.1	2.3	16.4	34.4	6.1	5.3	6.4	49.3
1974	9.2	2.5	16.4	33.9	5.4	5.1	5.6	50.6
1975	11.3	2.7	22.7	39.2	5.5	5.3	5.6	56.2
1976	11.8	3.1	20.8	39.0	6.1	5.7	6.3	56.8
1977	12.1	3.0	24.0	38.2	6.0	6.1	6.1	56.8
1978	11.4	2.7	21.3	35.7	6.2	5.9	6.3	54.4
1979	11.4	2.3	20.2	35.2	5.6	5.6	5.6	53.7
1980	11.1	2.3	19.7	32.6	5.6	5.8	5.5	49.5
1981	12.8	2.3	20.8	35.2	5.8	6.1	5.8	53.5
1982								
1983								
1984								

Figure A1.1 Tabulated data on DLO output based on Housing and Construction statistics.

year	age of building at end of year	structure	roofs	external finishes and claddings	external decoration	internal repairs	internal decoration	electrical services	plumbing services	heating services	external works	drainage	total maintenance
1962	6				603	18		46	11	72	20		770
1963	7			25		21		19	5	140	22		232
1964	8					30		28	7	168	209		442
1965	9		11	36		14		5	5	164	323	14	572
1966	10			5	217	62		22	19	35	119		479
1967	11		16	164		43	517	63	11	78	85		977
1968	12		11	10		30		9	14	62	82		218
1969	13		15	107		9			31	68	45		275
1970	14		4	94		74		12	25	151	101		461
TOTAL			57	441	820	301	517	204	128	938	1006	14	4426
average			6	49	91	33	57	23	14	104	112	2	492
percent			1.29	9.96	18.53	6.80	11.68	4.61	2.89	21.19	22.73	0.31	100.0

Figure A1.2 Study of maintenance costs for a school (£ at 1962 prices).

Tables of this nature can clearly become very complex, and simplified, rather more pictorial, techniques can be used. For example, block diagrams or pie charts are useful ways of describing the maintenance costs of a building, and pictograms are also convenient for illustrative purposes, such as that shown in figure A1.3, which gives the age profile of a housing association's housing stock.

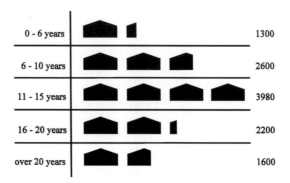

0 - 6 years	1300
6 - 10 years	2600
11 - 15 years	3980
16 - 20 years	2200
over 20 years	1600

Figure A1.3 Age composition of housing stock.

Frequency distributions

Much statistical data is concerned with the analysis of events or occurrences. For example, in maintenance operations, it may be necessary to study data on the frequency of the failure of components, or the frequency of call outs for emergency repairs.

Figure A1.4 shows the results of studying the failure time of a component for a group of 50 houses over a 15 year period. It is assumed, in this example, that the component has only failed once in every unit.

9	10	7	6	9	4	9	5	8	10
7	8	8	7	6	2	7	5	13	9
9	11	8	11	3	10	7	15	8	7
8	6	9	5	4	14	7	11	7	8
5	12	7	4	12	8	9	4	12	8

(a) Component failure years for 50 houses

year to failure	number of houses
1	0
2	1
3	1
4	4
5	4
6	3
7	9
8	9
9	7
10	3
11	3
12	3
13	1
14	1
15	1

(b) component failures grouped by years

year to failure	number of houses in band
1 - 3	2
4 - 6	11
7 - 9	25
10 - 12	9
13 - 15	3

(c) component failures grouped in year bands

Figure A1.4 Analysis of component failure times.

Table (a) is not very satisfactory, and a rearrangement as shown in (b)
is more informative. Alternatively, the data may be grouped, as shown
in (c), or presented graphically in any of the ways illustrated in figure
A1.5.

The failure time which appears in this data set varies continuously, i.e.
it is a continuous variable. The raw data is presented in a simplified form
as, in practice, few if any of these failures occur after a precise number of
years. The data only shows the years in which failure occurred.

Clearly a far more precise and complex set of data may be available,
giving the precise time of failure, e.g. 3.2 years. Time is an example of a

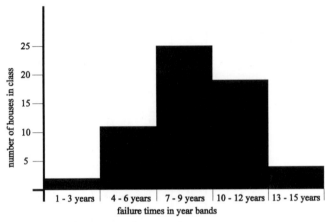

(a) Failure times as frequency histogram

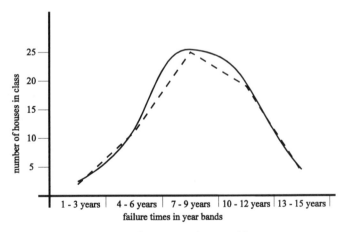

(b) Failure times as frequency polygon and frequency curve

Figure A1.5 Analysis of component failure times.

continuous variable but, for the purposes of representation, has been converted to a discrete variable. Other variables may already exist in a discrete form, such as the number of emergency monthly call-outs received by a maintenance depot over a 24 day working month, illustrated in figure A1.6. When continuous variables are being converted to discrete ones, some decision has to be made concerning the range or precision to be used.

MONTH 6				MONTH 7		
day	no. calls	% total		day	no. calls	% total
1	3	1.7		1	12	5.3
2	5	2.8		2	9	3.9
3	7	3.9		3	11	4.9
4	4	2.2		4	10	4.4
5	9	5.1		5	9	3.9
6	6	3.4		6	11	4.9
7	4	2.2		7	10	4.4
8	9	5.1		8	9	3.9
9	5	2.8		9	8	3.5
10	7	3.9		10	9	3.9
11	9	5.1		11	9	3.9
12	8	4.5		12	10	4.4
13	5	2.8		13	8	3.5
14	4	2.2		14	9	3.9
15	7	3.9		15	6	2.6
16	6	3.4		16	8	3.5
17	9	5.1		17	9	3.9
18	8	4.5		18	10	4.4
19	10	5.6		19	9	3.9
20	9	5.1		20	10	4.4
21	11	6.2		21	11	4.9
22	11	6.2		22	10	4.4
23	12	6.7		23	11	4.9
24	10	5.6		24	10	4.4

Figure A1.6 Tabulation of two months call out figures.

When presenting data, it is also important to be able to make simple comparisons. Figure A1.6 shows emergency calls for a maintenance depot in two separate months. A month-to-month analysis can be made by comparison of the actual call-out figures, or by their relative frequency, expressed in percentage terms. This can be represented graphically, by simple histograms or frequency polygons, as in figure A1.7.

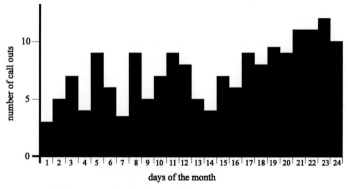

(a) Histogram of call-outs: month 6

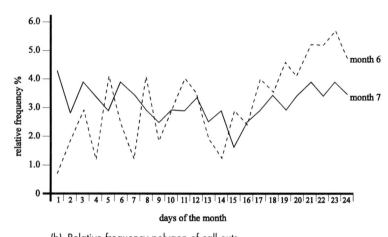

(b) Relative frequency polygon of call outs

Figure A1.7 Graphical representation of call out figures.

Statistics over time

A time series analysis studies a single variable quantity at intervals over a period of time. There are a number of simple examples of this in Chapter 1, where construction output data is represented in a simple graphical format. In all these cases intervals of a year are involved, and although expenditure is a continuous variable, it is discretised into annual sums. It may also be useful to study the maintenance costs for a building in a similar way, or to analyse annual maintenance costs by discretising costs on a monthly basis. A further step may be to examine relative monthly costs (figure A1.8).

	EXPENDITURE	% TOTAL
JANUARY	650	5.1
FEBRUARY	826	6.5
MARCH	745	5.8
APRIL	950	7.5
MAY	916	7.2
JUNE	1550	12.1
JULY	1675	13.1
AUGUST	1725	13.5
SEPTEMBER	1440	11.3
OCTOBER	950	7.5
NOVEMBER	721	5.7
DECEMBER	603	4.7
TOTAL	£12,751	100.0

Figure A1.8 Monthly breakdown of annual maintenance costs.

For carrying out comparisons, simple graphical representations are useful, but there is some value in the use of index numbers. These also have the additional merit of measuring the change in a variable with respect to a base point in time. Figure A1.9 compares the maintenance costs of two buildings over time in this manner.

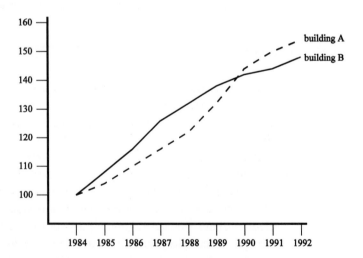

Figure A1.9 Indexed comparative annual maintenance costs at 1985 prices: 1984 base year = 100.

Cumulative representation

Statistical data, whether time related or not, as well as being presented in an incremental way, can also be presented in a cumulative fashion. Figure A1.10 shows the monthly expenditure from figure A1.9 represented in this way, and illustrates how this form of presentation gives a visual indication of the rate of expenditure from the slope of the polygon.

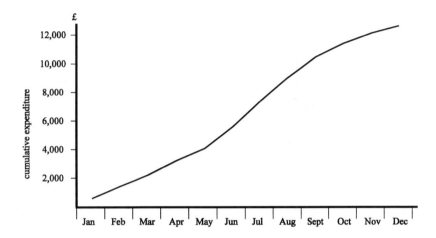

Figure A1.10 Monthly breakdown of annual maintenance costs.

The description of data

The presentation of data in a variety of ways allows the formulation of a qualitative view, from which inferences of a general nature may be drawn. To permit quantitative judgements, it is necessary to be able to characterise a data set by more objective means. If a frequency distribution, with grouping of data representing component lives, is available, an arithmetical mean can be determined as shown in figure A1.11.

Whilst the mean is the simplest and most widely understood measure it can often be a very misleading one, as extreme values can have a disproportionate effect. This is easily demonstrated in the following example. In carrying out a cost-in-use exercise, the following component lives in years are obtained from maintenance data, and the problem is one of deciding which should be used for evaluative purposes:

3, 3, 2, 9, 3, 9, 3, 2, 3, 2, 10, 3, 3, 2, 3

The average of these is slightly over 4, whereas 12 out of the 15 figures are less than this. In this simple case it may be decided to discard the

LIFE TO FAILURE	FREQUENCY F	MIDPOINT X	F x X
1 - 3	12	2	24
4 - 6	15	5	75
7 - 9	35	8	280
10 - 12	20	11	220
13 - 15	10	14	140

$\Sigma F = 92$ $\Sigma FX = 739$

AVERAGE LIFE $= \Sigma F / \Sigma FX$

$= 739/87$

$= 8.03$ YEARS

Note that this calculation assumes an even distribution within each class.

Figure A1.11 Calculation of mean component life.

figures which are obviously causing the distortion. However, for the purpose of analysing a data set objectively, there are other measures that can be used.

The median

If the above figures are arranged in order of magnitude:

2, 2, 2, 2, 3, 3, 3, **3**, 3, 3, 3, 3, 9, 9, 10

the median is the middle value in the series, i.e. 3.

Where larger amounts of data are involved, and perhaps grouped, the median can be found from the frequency polygon. Figure A1.12 shows how the median divides the area under the polygon in half, and it is possible to carry out a similar exercise to produce other divisions, and hence determine quartiles, or a variety of other percentiles. The cumulative frequency polygon may be easy to evaluate for these purposes.

The mode

The mode is another simple measure that may be more useful than an arithmetic average, and is the value which occurs most frequently in a series. In the simple example given above the mode is 3. For a larger data set it can very easily be determined from a frequency polygon (figure A1.13).

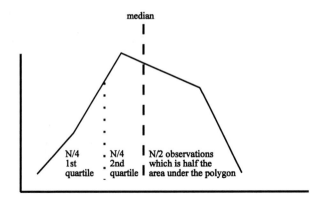

(a) Frequency polygon

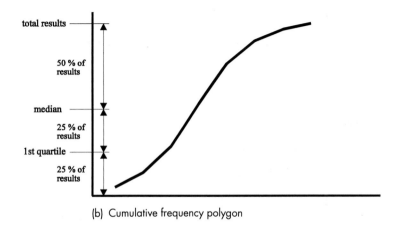

(b) Cumulative frequency polygon

Figure A1.12 Median and quartiles.

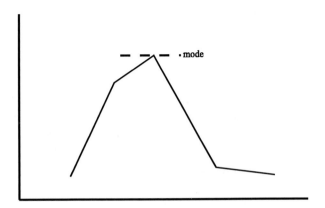

Figure A1.13 The mode from a frequency polygon.

Scatter, spread and dispersion

In analysing and comparing the number of daily call-outs for two different areas of an estate, the frequency polygons shown in A1.14 are produced. Analysis of the two sets of figures indicates that the average daily call-out number is roughly the same, yet their characteristic distributions are clearly different.

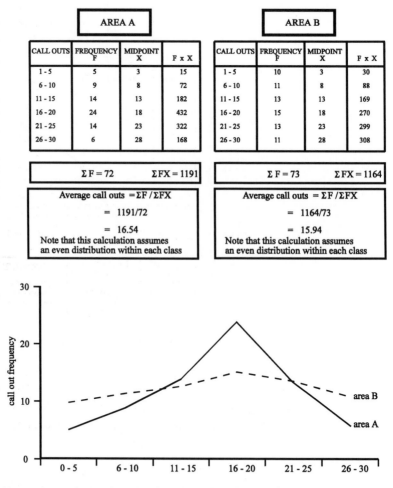

AREA A

CALL OUTS	FREQUENCY F	MIDPOINT X	F x X
1 - 5	5	3	15
6 - 10	9	8	72
11 - 15	14	13	182
16 - 20	24	18	432
21 - 25	14	23	322
26 - 30	6	28	168

$$\Sigma F = 72 \qquad \Sigma FX = 1191$$

Average call outs $= \Sigma F / \Sigma FX$

$= 1191/72$

$= 16.54$

Note that this calculation assumes an even distribution within each class

AREA B

CALL OUTS	FREQUENCY F	MIDPOINT X	F x X
1 - 5	10	3	30
6 - 10	11	8	88
11 - 15	13	13	169
16 - 20	15	18	270
21 - 25	13	23	299
26 - 30	11	28	308

$$\Sigma F = 73 \qquad \Sigma FX = 1164$$

Average call outs $= \Sigma F / \Sigma FX$

$= 1164/73$

$= 15.94$

Note that this calculation assumes an even distribution within each class

Figure A1.14 Comparative daily call-out figures for two areas over a three month period.

A measure of variability is needed to describe accurately the characteristics of a distribution. One way of doing this might be be to calculate the average deviation of each value in a data set from the mean. However, as figure A1.15 demonstrates this creates a problem. It is a

house number	maint. costs x	deviation from mean mean - x	deviation squared (mean - x)^2
1	198	15.6	243.36
2	186	27.6	761.76
3	209	4.6	21.16
4	206	7.6	57.76
5	211	2.6	6.76
6	196	17.6	309.76
7	230	-16.4	268.96
8	200	13.6	184.96
9	225	-11.4	129.96
10	218	- 4.4	19.36
11	240	-26.4	696.96
12	245	-31.4	985.96
13	200	13.6	184.96
14	212	1.6	2.56
15	248	-34.4	1183.36
16	228	-14.4	207.36
17	237	-23.4	547.56
18	170	43.6	1900.96
19	205	8.6	73.96
20	208	5.6	31.36

	total	Σ(mean - x)	Σ(mean - x)^2
	= £4272	= 0	= 7611.44

MEAN	STANDARD DEVIATION	VARIANCE
£4272/20 = £213.6	square root (£7611.4/20) = £19.50	£7611.4/20 = £380.57

Figure A1.15 Analysis of maintenance costs of a sample of 20 houses.

mathematical certainty that, if signs are taken into account, the positive and negative deviations will always cancel out. This can be overcome by ignoring the minus signs, which produces what is called a mean deviation. Of far greater benefit, however, is the standard deviation.

Standard deviation

The standard deviation can be calculated as shown in A1.15. Frequency curves are commonly used to describe graphically the characteristics of a data set, and the standard deviation. Natural phenomena, and events of chance, usually produce frequency curves which are bell-shaped, as shown in figure A1.16. In this example, the two curves represent data sets equal in size with the same arithmetic mean. The standard deviation, therefore, gives a means of distinguishing between the two.

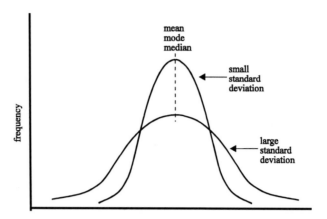

Figure A1.16 Comparative standard deviations.

Coefficient of variation

If two data sets have greatly different means, then the standard deviation is insufficient to distinguish them. The coefficient of variation expresses the standard deviation as a percentage of the mean.

Skewness

It is quite possible to have two sets of data that possess the same mean and the same standard deviation, but which have quite different distributions about the mean. The distinguishing factor between the two sets of data shown in figure A1.17 is the degree of symmetry, or skewness. Skewed curves are often characteristic of economic or social observations.

Another frequency curve is the so called bath-tub frequency curve, which is commonly used when considering the incidence of failure of building components through the life of a building.

The use of frequency distributions

The mathematical properties of frequency distributions can be described by an equation, in terms of a number of mathematical variables and constants including the following:

- ❑ The frequency of a given variable
- ❑ The total frequency of the distribution
- ❑ The standard deviation
- ❑ The mean of the distribution

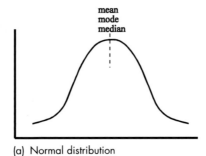

(a) Normal distribution

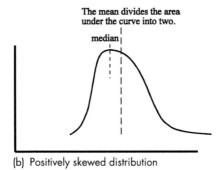

(b) Positively skewed distribution

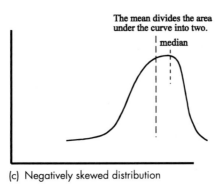

(c) Negatively skewed distribution

Figure A1.17 Comparative distributions.

For a normal distribution, using tables derived from such equations, it is possible to determine the proportion of cases between two vertical lines on the curve, i.e. the percentage of the total sample that has a value within a given range.

The principle is indicated in figure A1.18(a), with reference to what is called the standard normal distribution, which has a mean of zero and a standard deviation of one. For the general case, therefore, these percentages are in terms of multiples of the standard deviation (figure A1.18 (b)).

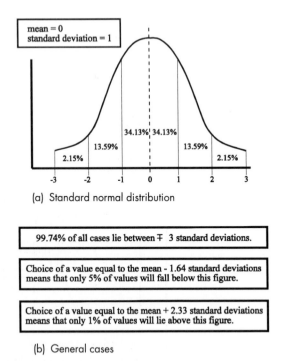

(a) Standard normal distribution

99.74% of all cases lie between ∓ 3 standard deviations.

Choice of a value equal to the mean - 1.64 standard deviations means that only 5% of values will fall below this figure.

Choice of a value equal to the mean + 2.33 standard deviations means that only 1% of values will lie above this figure.

(b) General cases

Figure A1.18 Analysis of the normal distribution.

Considerable manipulation is possible by using the tables, and this is best illustrated by an example. Figure A1.19 represents data on the failure profile of strip lights in a factory and, in instituting a planned replacement policy, it is necessary to consider a range of replacement cycles. This involves a management decision with respect to an acceptable chance of a failure between replacement cycles. In the example shown, this is carried out using the normal distribution. The figures are derived from mathematical tables, found in a number of good statistical texts, where a fuller exposition of the mathematical principles can also be found[5].

Similar exercises can be carried out on different distributions, but with a little more complexity. There are also a number of statistics computer packages that can be used, so that an intimate knowledge of theory is not necessary to make use of these techniques.

Probability and expected outcomes

Following the above exercise an evaluation of alternative strategies can be made, taking into account failure probabilities as shown in figure

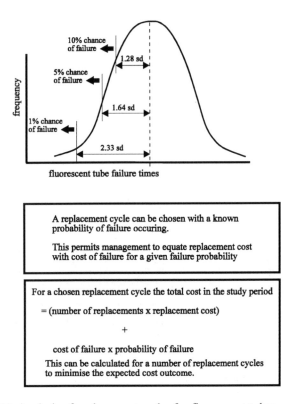

Figure A1.19 Analysis of replacement cycles for fluorescent tubes.

A1.19. It should be noted that a more rigorous analysis needs to take account of the time value of money. This introduces the notion of probability and fuller treatments are given elsewhere, along with related techniques used for decision making, such as decision trees.

Correlation techniques

Everyday experience throws up numerous instances of the manner in which different phenomena are associated in some way. For example, in Chapter 1 the relationship between maintenance output and a range of economic indicators was examined, and reference made to correlation. In its simplest form, this branch of statistics deals with two variables, and seeks to establish the relationship between them.

A local authority knows full well that there is a relationship between the age of its schools and annual maintenance costs, but needs to establish this in more objective terms. Figure A1.20 illustrates the results of the

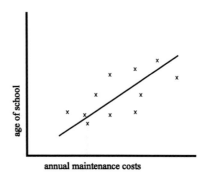

Figure A1.20 Scatter diagram for school maintenance costs.

authority plotting annual maintenance cost against age for its school stock in a particular area, and this is termed a scatter diagram. Assuming, rather loosely, that this relationship can be represented by a straight line, it is possible, using what are termed linear regression techniques, to determine the equation of the best straight line that fits this data, thus allowing extrapolation of maintenance costs into the future.

It is, of course, very debatable whether a straight line is appropriate, and it is possible to force a straight line on to any set of data. It is important to realise that this is only the best fit to the data under review. In figure A1.21 (a) there is an almost perfect fit whereas in (b) there is virtually none.

Such exercises, therefore, need to be qualified by other means. Before attempting the above exercise, it is wise to establish whether there is a numerical connection between the variables, and whether or not it is a linear one. The coefficient of correlation provides the measure of the degree of linear association present. It must be stressed that a poor correlation in this respect does not eliminate the possibility of a strong non-linear relationship, as shown in figure A1.22.

In some cases, normal direct plots may not show sensible linear correlation, but it should be borne in mind that alternatives are available such as a:

❑ Plot of one variable against the square of another
❑ Plot of one variable against the log of another

Such plots may produce a straight line even though a non-linear relationship exists.

Where a strong correlation is achieved, indicating a linear relationship, there will still be a degree of scatter. The outcome can be qualified, in such cases, by the statement of confidence limits as illustrated in figure A1.23[6].

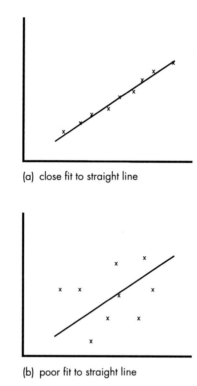

(a) close fit to straight line

(b) poor fit to straight line

Figure A1.21 Scatter diagram fit to straight line.

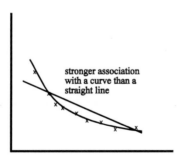

stronger association
with a curve than a
straight line

Figure A1.22 Curved versus straight line association.

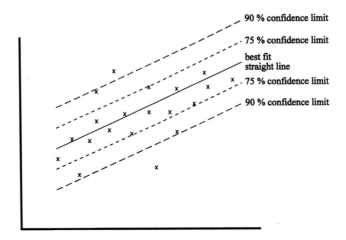

Figure A1.23 Confidence limits.

Time series analyses

Indexing

Indexing is a commonly used technique for measuring trends, particularly costs, over a period of time. For example, the retail price index and the number of indices used for measuring construction costs, are generally well known. The principle is demonstrated in figure A1.24, and the data could then be plotted on a graph. However, some care is needed when there are a number of contributory costs to be consolidated into one index.

Maintenance costs are made up of materials, labour and plant input, although the latter is often ignored. Indices are available, from a number

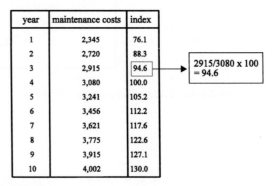

Figure A1.24 Indexing of maintenance costs.

of sources, for the individual inputs but, before they can be incorporated into a total maintenance cost index, their relative contribution has to be determined. This produces what is called a weighted index, and is the method used by BMI, and referred to in chapter 1. A simple example of such an exercise is illustrated in figure A1.25.

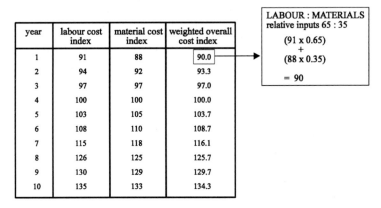

year	labour cost index	material cost index	weighted overall cost index
1	91	88	90.0
2	94	92	93.3
3	97	97	97.0
4	100	100	100.0
5	103	105	103.7
6	108	110	108.7
7	115	118	116.1
8	126	125	125.7
9	130	129	129.7
10	135	133	134.3

LABOUR : MATERIALS
relative inputs 65 : 35

(91×0.65)
$+$
(88×0.35)
$= 90$

Figure A1.25 Determination of weighted index costs.

Trends with time

If a variable, such as material costs, is plotted over time the true relationship may be masked somewhat by other variables. This means that the figures used may need to be adjusted to take account of such phenomena. For example, it might be expected that construction output would vary on a seasonal basis so that, when studying activity over a longer term, figures have to be seasonally adjusted.

In considering trends, the effects of random unforeseen events, such as damage caused by a storm of great magnitude that would only be expected to occur perhaps once every 100 years, need to be considered. If it happens to occur in the period under study, then clearly a distorted picture may emerge. There will also be cyclical variations that affect trends, if they are not in phase. The use of moving averages does much to eradicate some of these variables, and there are a number of other statistical techniques widely used[7].

References

(1)   Runyon, P.R. & Haber, A. (1971) *Fundamentals of Behavioural Statistics.* Addison-Wesley, Massachusetts.

(2) Bunting, J. (1995) *Mathematics and Statistics for the Built Environment*. Longman, London.
(3) Edelman, D.B. (1986) *Statistics for property people*. Estates Gazette, London.
(4) Sidwell, N.C. & Cheeseman, P.G. (1976) *Maintenance of CLASP Construction – Research Group Report*. Scottish Development Office.
(5) Runyon, P.R. & Haber, A. (1971) *Fundamentals of Behavioural Statistics*. Addison-Wesley, Massachusetts.
(6) Bunting, J. (1995) *Mathematics and Statistics for the Built Environment*. Longman, London.
(7) Bunting, J. (1995) *Mathematics and Statistics for the Built Environment*. Longman, London.

Appendix 2

Methods of Financial Appraisal

The value of money over time

There are a number of financial appraisal techniques, ranging from the simple to the sophisticated, that can be of use as an aid to decision making in many areas of building design and evaluation, as well as during the management of the building in use. The simplest technique, the straightforward payback approach, suffers from a failure to take into account the time value of money. Offered £1000 today, or the same amount in a year's time, it will be beneficial to take the money now, on the basis that it can be invested, and therefore be worth more in the future.

At an interest rate of 10%, £1000 invested today will be worth £1100 in a year's time. Viewed conversely it can be said that £1100 in a year's time has a present value of £1000.

£1000 to be received in a year's time has a present value of:

$$£1000 \times \frac{100}{110} = £909.10$$

In mathematical terms, the present value (PV) of £1 after n years with an interest rate of $i = 1/(1+i)^n$. In practice, tables of discount factors are available, to enable future cash flows to be discounted to a present value. Figure A2.1 shows an extract from such a table.

PRESENT VALUE OF £1

no. of years	interest	
	6%	9%
5	0.7473	0.6499
10	0.5584	0.4224
20	0.3118	0.1784
30	0.0303	0.0057

Figure A2.1 Present value (PV) of £1 in the future.

Thus, if the cost of replacing a component in 10 years' time is £1500 then the present values of replacement are:

PV = £1500 × 0.5584 = £837.60 @ 6% discount rate
PV = £1500 × 0.4224 = £633.60 @ 9% discount rate

Many costs are recurring annual ones and, whilst each year could be discounted separately, there is also a formula, and corresponding table, that facilitates the determination of the present value of a regular series of cash flows over a period of years. An extract of a table is shown in figure A2.2.

PRESENT VALUE OF £1 PER ANNUM

no. of years	interest	
	6%	9%
5	4.212	3.890
10	7.360	6.418
20	11.470	9.129
30	13.765	10.274
60	16.161	11.048

Figure A2.2 Present value (PV) of £1 per annum in the future.

Thus, if there is an annual cleaning cost, over a presumed 60 year building life, of £1500 per annum, the respective present values are:

PV = £1500 × 16.161 = £24 241.50 @ 6% discount rate
PV = £1500 × 11.048 = £16 572.00 @ 9% discount rate

There are a number of variations on the same theme, all based around the same mathematical principles.

Net present values

The net present value technique (NPV) involves discounting all future cash flows to a common base year. Suppose that a design team is faced with the following scenario:

There are two alternative ways of providing the cladding to a building. Alternative A involves an initial outlay of £100 000 with predicted annual maintenance of £2000. Alternative B has an initial outlay of £120 000 with annual predicted maintenance costs of £500. The cash flow, and discounted equivalents, at a rate of 9%, are shown in figure A2.3.

Whilst, in simple cash flow terms, option A appears to be the most expensive option, when the figures are discounted it has the lower

ALTERNATIVE A

cash flow	year(s)	discount factor	present value
£100 000	1	1	£100 000
£ 2 000 (x 60 yrs)	1 - 60	11.048	£ 22 096
£220 000			£122 096

ALTERNATIVE B

cash flow	year(s)	discount factor	present value
£120 000	1	1	£120 000
£ 500 (x 60 yrs)	1 - 60	11.048	£ 5 524
£150 000			£125 524

Figure A2.3 Comparison of alternatives using present values.

present value. This can be put another way. An investment now of £122 096 will pay the total cost of alternative A over its design life, but an equivalent investment now of £125 524 will be required to pay for option B.

Annual equivalent

Another approach that can be taken, based on the same principles, is the so called annual equivalent method. In this technique, the cash flows throughout the life of an asset are converted into an equivalent annual cost. In terms of evaluation, it will rank alternatives in exactly the same order as the net present value approach, but will present the figures in a more meaningful way for the building owner.

To obtain the annual equivalent, the NPV is divided by the PV of £1 per annum for the appropriate number of years. For example to convert the PV for alternative A above to an annual equivalent, £122 096 divided by 11.048 gives £11 051. In other words, an expenditure of £11 051 per year over 60 years has a present value of £122 096.

The usefulness of the technique for life cycle costing is shown in figure A2.4, where a series of cash flows over time, that are not recurring in-phase can be converted to an equivalent annual expenditure.

Softwood windows require a capital outlay of £4 500 and renewal every 15 years at a cost of £5 000 at today's prices, including an allowance for removing the old ones. Redecoration will be required every five years at a cost of £400. The economic life of the building is taken to be 60 years.
The annual equivalent cost is determined below.

cash flow	year(s)	discount factor	present value
£4 500	0	1.0000	£4 500
£ 400	5	0.6499	£ 260
£ 400	10	0.4224	£ 169
£5 000	15	0.2745	£1 373
£ 400	20	0.1784	£ 71
£ 400	25	0.1160	£ 46
£5 000	30	0.0754	£ 377
£ 400	35	0.0490	£ 20
£ 400	40	0.0318	£ 13
£5 000	45	0.0207	£ 104
£ 400	50	0.0134	£ 5
£ 400	55	0.0087	£ 3

total present value　£6 941

60 year £1 per annum discount factor 11.048

Annual Equivalent = £6 941/11.048

annual equivalent　£ 628

Figure A2.4 Determination of annual equivalent over 60 year life.

Internal rate of return

Another derivative technique is the determination of a so-called internal rate of return. The object of the exercise, in this case, is to determine the interest rate that will produce, when all future cash flows, positive and negative, are taken into account, a net present value of zero. That is, when discounted costs equate to discounted benefits.

Determination of the internal rate of return is most simply carried out by determining the NPV of a set of cash flows at various discount rates and plotting them onto a graph, from which the IRR can be found.

Sinking funds

When allowance has to be made to meet a known future capital expenditure, one of the more prudent ways of doing this is to set up a sinking fund. This involves setting aside a regular sum of money that, when invested, will accumulate sufficiently to meet that future commitment. A good example is the requirement for housing associations to allow for major renewal programmes.

The requirement in this case is to determine the amount of money that needs to be set aside annually, at a given discount rate, to amount to the capital requirement in a number of years' time.

An industrial organisation predicts that it needs to carry out a major refurbishment to a production unit in five years' time that will cost £100 000. Determine the amount it needs to invest in a sinking fund to achieve this, assuming a discount rate of 8%. Discount tables provide a uniform series sinking fund factor. For five years at 8% this is 0.17045.

The required amount per year is therefore: £100 000 × .17045 = £17 045.

In other words, the setting aside of £17 045 each year for the next five years will accumulate to £100 000 at an interest rate of 8%.

Index